Raman Spectroscopy - Advances and Research Frontiers

Edited by Touseef Para

Published in London, United Kingdom

Raman Spectroscopy - Advances and Research Frontiers
http://dx.doi.org/10.5772/intechopen.1009037
Edited by Touseef Para

Contributors
Affi Nur Hidayah, Diana Podoleanu, Elena Starodub, Gheorghe Chirman, Ion Munteanu, Nicolae A. Enaki, Olena Zavorotynska, Seiji Kojima, Takuji B. M. Adachi, Tatiana Paslari, Touseef Para, Xiaoyu Zhao, Zihan Li

First published in London, United Kingdom, 2025 by IntechOpen
IntechOpen is the global imprint of INTECHOPEN LIMITED, registered in England and Wales, registration number: 11086078, 167-169 Great Portland Street, London, W1W 5PF, United Kingdom

For EU product safety concerns: IN TECH d.o.o., Prolaz Marije Krucifikse Kozulić 3, 51000 Rijeka, Croatia, info@intechopen.com or visit our website at intechopen.com.

British Library Cataloguing-in-Publication Data
A catalogue record for this book is available from the British Library

Raman Spectroscopy - Advances and Research Frontiers
Edited by Touseef Para
p. cm.
Print ISBN 978-1-83635-674-5
Online ISBN 978-1-83635-673-8
eBook (PDF) ISBN 978-1-83635-675-2

If disposing of this product, please recycle the paper responsibly.

Meet the editor

Dr. Touseef Para holds MSc, MPhil, and Ph.D. degrees in Physics with a specialization in Materials Science. His current research focuses on energy storage and conversion, with experience spanning solar cells, sodium-ion batteries, and related instrumentation. In addition to his research contributions, he has worked extensively with Raman spectroscopy and other characterization techniques applied to advanced materials. As editor of *Raman Spectroscopy – Advances and Research Frontiers*, he brings together contributions from leading experts to present current developments and emerging directions in the field.

Contents

Preface

Raman spectroscopy has established itself as a central technique for probing the vibrational and structural properties of matter. Since its discovery, the method has advanced significantly in instrumentation, sensitivity, and the range of applications. Today, it is widely used not only in fundamental research but also in applied fields such as materials science, energy research, and the life sciences.

This book, *Raman Spectroscopy – Advances and Research Frontiers*, presents recent developments in both methodological aspects and application areas. The book begins with an introductory section, followed by two main sections.

The first section provides a brief, concise introduction to spectroscopy and Raman scattering, while introducing both classical and quantum-mechanical understanding.

The second section presents progress in instrumentation and methodology. It includes chapters on improvements in signal generation and collection in Raman spectroscopy, the use of terahertz Raman spectroscopy for investigating ferroelectric materials, and the influence of nanoparticle morphology on plasmonic and polaritonic effects that enhance Raman sensitivity.

The third section is devoted to applications. It covers the role of Raman spectroscopy in the study of hydrogen storage materials, advances in three-dimensional Raman imaging for biomedical investigations, and the analysis of coherent Raman processes in biomolecular complexes.

Together, these contributions provide an overview of ongoing work at the interface of technique development and practical application. The book is intended for researchers and students who seek to follow current directions in Raman spectroscopy and to apply these methods in their own investigations.

I am thankful to all contributors for their efforts in preparing the chapters and hope that the book will serve as a valuable reference for the community. I am also thankful to IntechOpen for providing a platform where knowledge is not restricted behind firewalls and payment gateways.

Touseef Para
Rechargion Energy,
Pune, India

Section 1

Introduction

Chapter 1

Introductory Chapter: Raman Spectroscopy – Advances and Research Frontiers

Touseef Para

1. Spectroscopy

Spectroscopy is the study of interactions between electromagnetic radiation and matter. These interactions can unveil the structure, composition, and dynamics of molecules, atoms, and condensed phases. By examining absorption, emission, and scattering processes, it allows identification of energy levels, molecular vibrations, and electronic transition/states. Techniques such as infrared spectroscopy reveals molecular vibrations and functional groups, ultraviolet-visible spectroscopy provides information on electronic transitions, and nuclear magnetic resonance spectroscopy gives insights into molecular structure and chemical environment. X-ray diffraction and photoelectron spectroscopy are used to examine crystal structures and electronic states in solids. Among these, Raman spectroscopy is a powerful and non-destructive technique that provides detailed information on molecular vibrations and lattice dynamics without requiring extensive sample preparation. It can be applied to solids, liquids, and gases and is especially useful for studying materials in their natural state. It is therefore the purpose of this book to introduce readers to the current situation of research in this area. This chapter introduces the general principles of spectroscopy, reviews its main types, and develops the theoretical framework for Raman scattering.

2. Fundamentals of spectroscopy

2.1 Electromagnetic radiation and matter

Electromagnetic radiation is characterized by its wavelength λ or frequency ν, related by

$$\nu = \frac{c}{\lambda} \tag{1}$$

where c is the speed of light. The energy of a photon is expressed as

$$E = h\nu \tag{2}$$

with h being Planck's constant. When radiation interacts with matter, transitions may occur between discrete energy levels, which can be electronic, vibrational, rotational, or spin-related, depending on the energy of the radiation.

2.2 Outcomes of light-matter interaction

The interaction of photons with matter can result in the following:

1. *Absorption*—energy from a photon is transferred to the system, promoting it to a higher energy state.

2. *Emission*—the system releases energy as photons while returning to a lower energy state.

3. *Scattering*—photons are deflected by the system, either elastically (Rayleigh) or inelastically (Raman or Brillouin).

3. Classification of spectroscopic techniques

3.1 Absorption spectroscopy

- UV-Visible (UV-Vis): probes electronic transitions in molecules and materials.

- Infrared (IR): probes vibrational transitions, providing a molecular fingerprint.

3.2 Emission spectroscopy

- Flame/Plasma emission: elemental analysis through characteristic atomic emission lines.

- Photoluminescence (PL): emission from electronically excited states, used in semiconductors and materials research.

3.3 Scattering-based spectroscopy

- Rayleigh scattering: elastic scattering with no change in photon energy.

- Raman scattering: inelastic scattering, provides information on vibrational and rotational modes.

- Brillouin scattering: inelastic scattering from acoustic phonons, used to determine elastic properties.

3.4 Magnetic resonance spectroscopy

- Nuclear magnetic resonance (NMR): interaction of nuclear spins with radiofrequency radiation, used for structural determination.

- Electron spin resonance (ESR/EPR): probes unpaired electron spins, used in materials and biochemical studies.

4. Raman spectroscopy

4.1 Historical context

Raman spectroscopy was first discovered by C. V. Raman and K. S. Krishnan in 1928 [1], when they observed the inelastic scattering of light from liquids, leading to a shift in wavelength known today as the Raman effect. This discovery provided experimental evidence that light can exchange energy with molecular vibrations, complementing the understanding of molecular spectra obtained through infrared absorption and fluorescence. The pioneering experiments by C. V. Raman utilized monochromatic light sources and simple spectrographs to detect weak scattered radiation, demonstrating the existence of Stokes and anti-Stokes lines. For this fundamental contribution, C. V. Raman was awarded the Nobel Prize in Physics in 1930, establishing Raman spectroscopy as a powerful tool for probing molecular structure and vibrational dynamics. Early studies laid the groundwork for subsequent theoretical developments, including classical and quantum descriptions of scattering processes, which remain central to the interpretation of Raman spectra today [2].

4.2 Basic principle

When light interacts with molecules, most photons scatter elastically (Rayleigh scattering), while a small fraction undergo inelastic scattering. In inelastic scattering, photons exchange energy with molecular vibrations, leading to a shift in photon frequency [3].

- Stokes scattering: photon loses energy ($\nu_s < \nu_0$).
- Anti-Stokes scattering: photon gains energy ($\nu_s > \nu_0$).

The Raman shift is defined as:

$$\Delta\nu = \nu_0 - \nu_s \tag{3}$$

where ν_0 is the frequency of incident light and ν_s is the frequency of scattered light.

4.3 Classical theory

In the classical description [4, 5], a molecule is treated as a polarizable entity. When an oscillating electric field $\mathbf{E} = \mathbf{E}_0 \cos(2\pi\nu_0 t)$ from incident light interacts with the molecule, it induces a dipole moment $\mathbf{P}$ given by:

$$\mathbf{P} = \boldsymbol{\alpha} \cdot \mathbf{E} \tag{4}$$

where $\boldsymbol{\alpha}$ is the polarizability tensor. For a vibrating molecule, the polarizability depends on the normal coordinates Q_k of vibration:

$$\alpha_{ij} = (\alpha_{ij})_0 + \sum_k \left(\frac{\partial \alpha_{ij}}{\partial Q_k}\right)_0 Q_k. \tag{5}$$

Assuming harmonic vibrations $Q_k = Q_{0k} \cos(2\pi\nu_k t)$, the induced dipole moment contains components oscillating at frequencies ν_0, $\nu_0 + \nu_k$, and $\nu_0 - \nu_k$, corresponding to Rayleigh, anti-Stokes, and Stokes scattering, respectively.

The classical criterion for Raman activity requires that the derivative of the polarizability tensor with respect to the vibrational coordinate is nonzero:

$$\frac{\partial \alpha_{ij}}{\partial Q_k} \neq 0. \tag{6}$$

This framework explains the origin of Raman shifts and provides a basis for understanding the frequency dependence of scattered light intensity.

4.4 Quantum mechanical description

In the quantum mechanical framework [6, 7], Raman scattering is treated as a second-order perturbation process involving virtual electronic states. A molecule initially in a vibrational state $|i\rangle$ of the electronic ground state absorbs a photon and is momentarily promoted to a virtual electronic state $|v\rangle$, followed by emission of a photon that leaves the molecule in a final vibrational state $|f\rangle$. The Raman scattering amplitude is determined by the transition polarizability tensor, given by

$$\alpha_{fi} = \sum_v \left[\frac{\langle f|\mu|v\rangle \langle v|\mu|i\rangle}{E_v - E_i - h\nu_0 + i\Gamma_v} + \frac{\langle f|\mu|v\rangle \langle v|\mu|i\rangle}{E_v - E_i + h\nu_s + i\Gamma_v} \right], \tag{7}$$

where μ is the dipole moment operator, E_v and E_i are the energies of the virtual and initial states, $h\nu_0$ and $h\nu_s$ are the energies of the incident and scattered photons, and Γ_v accounts for the lifetime broadening of the virtual state. The intensity of Raman scattering is proportional to the square of the transition polarizability and scales with the fourth power of the scattered frequency. This approach naturally explains both Stokes and anti-Stokes processes and provides a foundation for understanding resonance effects and the polarization dependence of Raman scattering [8].

The Intensity of Raman line can be expressed as

$$I_{fi} \propto (\nu_0 - \nu_k)^4 |\alpha_{fi}|^2 \tag{8}$$

The intensity is proportional to the square of the transition polarizability and depends on the frequency of the scattered light.

For a vibrational mode k, the intensity of the Stokes line can be expressed as

$$I_k \propto (\nu_0 - \nu_k)^4 |\alpha_k|^2 \frac{1}{1 - e^{-h\nu_k/k_B T}} \tag{9}$$

where ν_0 is the frequency of the incident light, ν_k is the frequency of the vibrational mode, α_k is the component of the polarizability tensor associated with the mode, h is Planck's constant, k_B is Boltzmann's constant, and T is the absolute temperature.

The factor $(\nu_0 - \nu_k)^4$ arises from the classical electromagnetic theory, reflecting the dependence of scattering intensity on the frequency of scattered radiation. The term $1/\left(1 - e^{-h\nu_k/k_B T}\right)$ accounts for the thermal population of vibrational levels and differentiates Stokes and anti-Stokes intensities [9].

4.5 Selection rules

Selection rules in Raman spectroscopy [10] dictate which vibrational modes of a molecule are Raman active. Unlike infrared absorption, which requires a change in the dipole moment, Raman activity arises from a change in the polarizability tensor during vibration. Mathematically, a vibrational mode is Raman active if the derivative of at least one component of the polarizability tensor with respect to the vibrational coordinate is nonzero, that is, $\frac{\partial \alpha_{ij}}{\partial Q_k} \neq 0$. Group theory provides a systematic method to predict Raman activity by analyzing the symmetry properties of the vibrational modes and their corresponding irreducible representations in the point group of the molecule. These representations determine whether a mode transforms like the quadratic functions $(x^2, y^2, z^2, xy, yz, zx)$ associated with polarizability. As a result, Raman selection rules complement infrared selection rules, making the two techniques jointly powerful for probing molecular structure.

4.6 Symmetry considerations

Symmetry rules in Raman spectroscopy [10] are derived from the molecular point group and govern the activity of vibrational modes based on the transformation properties of the polarizability tensor. Each vibrational mode belongs to a specific irreducible representation of the molecule's point group. A mode is Raman active if the components of the polarizability tensor transform according to the same irreducible representation as the vibrational mode. Group theory provides a systematic framework to determine which modes satisfy this criterion, allowing prediction of the number and type of Raman-active vibrations. These symmetry considerations not only determine whether a mode appears in the Raman spectrum but also influence the polarization properties of the scattered light, which can be experimentally exploited to gain additional structural information about the molecule or crystal. Symmetry plays a crucial role in determining the Raman activity of vibrational modes. The molecular point group dictates the behavior of the polarizability tensor under symmetry operations.

4.7 Resonance Raman spectroscopy

In resonance Raman spectroscopy [7], the frequency of the incident radiation is close to an electronic transition of the molecule, resulting in a significant enhancement of the Raman scattering cross-section for vibrational modes coupled to that electronic transition. This selective enhancement allows detection of weak or otherwise inactive vibrational modes and provides detailed information about the electronic structure and vibronic coupling. The resonance effect arises because the intermediate virtual state in the Raman process coincides with a real electronic excited state, increasing the probability of scattering. Resonance Raman spectroscopy is particularly useful for studying chromophores in biological molecules, conjugated systems in materials science, and transition-metal complexes, enabling selective probing of specific functional groups and structural features that are difficult to resolve in conventional Raman spectroscopy. In resonance Raman spectroscopy, the incident light frequency is close to an electronic transition of the molecule. This resonance condition enhances the Raman scattering cross-section, allowing for more sensitive detection of specific vibrational modes.

5. Conclusion

Spectroscopy provides fundamental insights into the structure and properties of matter. Raman spectroscopy, through the analysis of inelastic scattering of light, offers detailed vibrational information complementary to other techniques. The theoretical foundations outlined here establish the basis for understanding and interpreting Raman spectra.

The sections on instrumentation and applications presented here provide an overview of the principles and general uses of Raman spectroscopy. These topics are explored in greater detail in the subsequent chapters.

Author details

Touseef Para[1,2]

1 DESE, IIT Bombay, Mumbai, India

2 Rechargion Energy, Pune, India

*Address all correspondence to: drtouseefpara@yahoo.com

References

[1] Raman CV, Krishnan KS. A new type of secondary radiation. Nature. 1928;**121**: 619

[2] Raman CV, Krishnan KS. The optical analogue of the Compton effect. Nature. 1928;**122**:3074

[3] Long DA. Raman Spectroscopy. New York: McGraw-Hill; 1977

[4] Herzberg G. Molecular Spectra and Molecular Structure. Vol. 2. New York: Van Nostrand; 1945

[5] Kramers HA, Heisenberg W. Über die streuung von licht an licht. Zeitschrift für Physik. 1925;**31**:681-708

[6] Long DA. Early history of the Raman effect. International Reviews in Physical Chemistry. 1988;**8**(1):1-26

[7] Schmidt MK et al. Quantum mechanical description of Raman scattering from plasmon-enhanced nanostructures. ACS Nano. 2016;**10**(6): 5761-5770

[8] Kranert C et al. Raman tensor formalism for optically anisotropic crystals. Physical Review Letters. 2016; **116**:127401

[9] Keresztury G. Raman Spectroscopy: Theory. New York: Wiley; 2002

[10] Raman CV. A change of wave-length in light scattering. Nature. 1928;**121**:619

Section 2

Instrumentation

Chapter 2

Raman Microspectroscopy: Improvement in Signal Generation and Collection to Facilitate Raman Spectroscopy

Zihan Li and Takuji B.M. Adachi

Abstract

Raman microspectroscopy offers several advantages over a traditional Raman spectroscopy setup. The major one is its superior signal collection ability owing to the use of high numerical aperture objective lens. The choice of optimal pinholes and high-quality filters also allows one to eliminate unwanted background signals to the least. Such a setup enables one to use a high-power continuous-wave (CW) laser with GHz bandwidth, not MHz, to acquire high-quality Raman spectrum at room temperature. The acquisition of Raman spectrum can therefore be performed with much less acquisition time at a proper home-built Raman microspectroscopy setup. This opens up various applications of Raman spectroscopy toward *in situ* spectroscopy measurement to identify crystal polymorphs, time-resolved Raman spectroscopy to follow the phase transition of materials, the detection of inorganic nanocrystals in plastic products, *etc*. Herein, we describe the key practical points to consider in designing and building a setup, and then introduce several applications that have been recently published from our group.

Keywords: Raman microspectroscopy, confocal microscopy, *in situ* measurement, time-resolved Raman spectroscopy, *in vivo* characterization

1. Introduction

Raman spectroscopy is a powerful vibrational spectroscopic technique widely used for material identification, molecular structure analysis, and chemical composition profiling across various fields such as chemistry, biology, pharmaceuticals, and materials science [1–6]. Its label-free and non-destructive aspect facilitates its application to a wide range of systems. A vibrational mode appears in a Raman spectrum when it involves a change in the polarizability of molecules. As the frequency of each vibration depends on bond strength and the mass of atoms involved in the vibration, Raman spectroscopy has been used routinely as a molecular characterization tool [7–10]. Recently, owing to the development of optical filters, even the THz region of Raman spectrum that corresponds to lattice phonon modes has become readily accessible [11–13].

Despite the rich information Raman spectroscopy offers, Raman scattering is an inefficient physical process, and the signal is weak. The cross section of Raman scattering is typically in the range of 10^{-30} cm^2, ~14 orders of magnitude smaller than that of photoabsorption. Thus, Raman spectral measurements can be seen as a slow spectroscopy method that requires a long acquisition time. One way to overcome this limitation has been the use of coherent Raman spectroscopy such as Coherent anti-Stokes Raman spectroscopy (CARS) and stimulated Raman scattering (SRS) spectroscopy, which utilizes a combination of ultrafast laser pulses to amplify Raman signals [14, 15]. Another simpler alternative is its implementation in an optical microscopic configuration. When a laser beam is focused to a diffraction-limited spot using a high numerical aperture (NA) objective lens (*cf.* millimeter-sized beam used in conventional Raman spectroscopy), one can improve the signal-to-background ratio and therefore shorten the measurement time required to obtain a good quality spectrum. The improvement is particularly important for analyzing a system with micro- or nanoscale materials. As the focused beam spot is typically below 1 μm, it can also be used to scan through a sample in 2D or 3D. Indeed, lately, there are many commercial confocal Raman microscopes on the market (typically costs >300 k euros). Although a commercial instrument offers a user-friendly software and it is easy to use even for beginners, it is basically a black box that its mechanics can no longer be accessed by users. A home-built setup offers flexibility in tuning experiments (type of laser source, laser power and wavelength, the polarization state of a laser, *etc.*), and this can be a crucial asset for a laboratory to perform the exact experiments as desired.

In this chapter, our goal is to share practical know-how on designing and building a setup, with a strong emphasis on the aspect of superior signal collection of Raman microspectroscopy rather than imaging capability. When the management of discriminating signal from background is properly executed, the setup can perform spectroscopy at time resolution relevant for *in situ* and real-time analysis, which is particularly valuable for monitoring dynamic biological processes, material responses to stress, and ongoing chemical reactions [16–19]. The combination of high spatial and temporal resolution makes it possible to be applied in a broader area of applications that were not accessible by conventional Raman spectroscopy. For example, it can detect the appearance of various crystal polymorphs *in situ,* including transient metastable phases [20], perform real-time monitoring of phase transitions [21], and detect nanoscale inorganic particles in composite materials such as plastics [4]. Following the critical design considerations for building an effective Raman microspectroscopy system, we will discuss several representative applications performed on such a setup.

2. Construction of Raman microspectroscopy setup

The instrumentation of a Raman microscope typically includes an excitation laser, a microscope body (consisting of dichroic mirrors, microscope objective, tube lens, *etc.*), optical filters, and detectors. **Figure 1** shows a Raman microspectroscopy setup built in our group [22]. A commercial microscope body can be used if users prefer to build a setup around it, because it facilitates the alignment process of the laser beam. One can, of course, construct a microscope body on their own if they prefer more flexibility or need to reduce the cost. The choice of an upright or inverted microscope

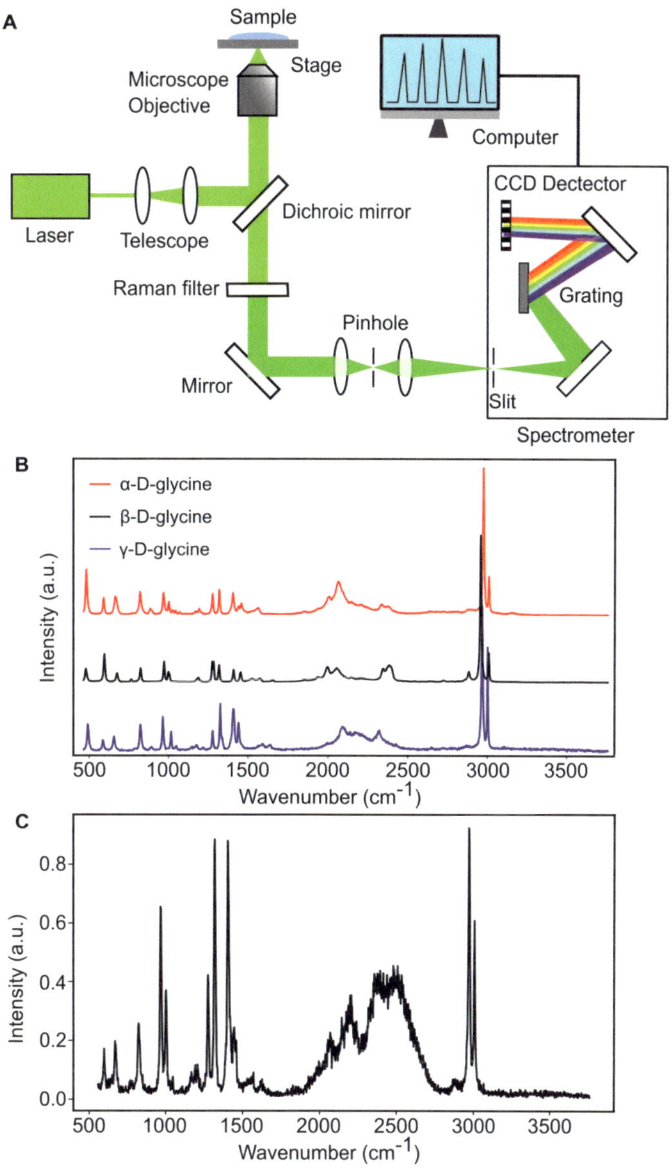

Figure 1.
(A) Schematic representation of a home-built Raman microspectroscopy setup. (B) Overlaid Raman spectra of three deuterated glycine crystal polymorphs. 532 nm laser (Quantum Design, Opus 532) was focused on a sample using 60 × water immersion objective (NA 1.2). The laser power was 150 mW at the sample. Spectra were acquired with 10 s exposure, accumulated over 10 frames. (C) Raman spectrum of α-D-glycine measured on the same setup with the laser power of 1.2 W at the sample. The spectrum was obtained by a single frame acquisition with 20 ms exposure time.

is also up to the preference of users. We tend to choose an inverted microscope design because of the sample configurations we often use.

In principle, a laser beam is collimated and introduced to a microscope objective to focus on a sample. The diameter of the beam must be adjusted to overfill the back aperture of an objective lens. A telescope (Keplerian or Galilean design) with an optimal magnification is used for this purpose. The collection of Raman scattering signal is done by the same objective lens (*epi* configuration). Raman edge filters are used to reject Rayleigh scattering (long-pass type dichroic mirror and long-pass filter). Then, the signal is detected by a spectrometer equipped with a high-sensitivity charge-coupled device (CCD). The numerical aperture (NA) of an objective lens is a key parameter that determines the size of laser focus at a sample and the power of signal collection. It is recommended to have several lenses with a range of NA, so that one can adjust the spot size depending on the size of the sample structural feature or circumvent the risk of photodamages. The quality of optical filters is important to minimize the amount of Raman signals to be blocked by the filter near the cutoff wavelength. In addition to the main components briefly described, a pinhole is usually placed to improve the discrimination between signal and background [23]. It plays a crucial role in minimizing the detection of background noise and stray light, thus significantly improving the signal-to-background ratio of the collected spectra. Such a setup supports the use of high-power continuous-wave (CW) lasers with the linewidth in the GHz range, which enables faster acquisitions of Raman spectra without compromising the spectral quality at room temperature.

2.1 Main components of a Raman microspectroscopy setup

2.1.1 Laser

In Raman microspectroscopy, the choice of laser source is critical, as it directly influences Raman scattering intensity, spatial resolution, potential fluorescence background, and acquisition time. Here, we do not discuss fundamental physics on how the excitation wavelength affects Raman scattering and spectrum (*e.g.*, resonance Raman scattering). We will mainly focus on introducing different types of lasers and practical aspects of how to select one.

Traditionally, gas lasers such as Ar ion and He-Ne lasers were the popular choices for Raman spectroscopy. Owing to technological developments, single-frequency lasers and diode-pumped solid-state lasers (DPSSLs) are more commonly used nowadays. Single-frequency lasers, available with a wide range of emission wavelengths, are compact with power outputs typically ranging around 100 mW [24]. Single-frequency lasers are more suited for Raman spectroscopy applications because of their narrow spectral bandwidth, which is typically in the range of MHz. These are the lasers typically adopted in many commercial confocal Raman microscopes. DPSSLs, including frequency-doubled Nd:YAG or Nd:YVO_4 lasers emitting at 532 or 1064 nm, offer excellent beam quality, single longitudinal mode output. While their laser linewidth ~30 GHz is not as narrow as single-frequency lasers [25], these lasers typically deliver much higher power up to 20 W. Despite the GHz range linewidth, it delivers high enough quality performance for most of Raman spectroscopy at room temperature. The advantage of this class of lasers is their power that could simply increase the Raman scattering intensity, as long as the potential issues such as photodamage or sample heating can be managed. Tunable lasers, such as a CW tunable Ti:Sapphire laser (e.g., 650–1100 nm), have not been commonly used because the wavelength is not locked. Although the linewidth is also GHz range and therefore it is usable for most room temperature Raman spectroscopy, there are non-negligible amount of unwanted side-bands for spectroscopy. If one desires to use such lasers for Raman spectroscopy, one must characterize the wavelength of the laser frequently using a wavemeter (scanning Michelson interferometer). In addition, laser line filter is needed to clean up the side-bands to ensure they will not contaminate the spectra. Once these points are taken care of, adjustable excitation wavelengths provide users with advantages for the application of resonance Raman spectroscopy or minimizing fluorescence background. For many applications at room temperature, users have several options of lasers. The selection of laser type depends on the desired specification of spectroscopy, such as spectral resolution, laser power, and time resolution needed for specific Raman applications.

2.1.2 Objective lens

The objective lens plays a critical role in both focusing the excitation laser onto the sample and collecting the scattered Raman signal. One of the most important parameters is the numerical aperture (NA), the ability of lens to gather light. Higher NA objectives allow a tighter focusing and more efficient light collection, leading to improved spatial resolution and better signal-to-background ratio. Depending on the experimental requirements and sample types, a suitable objective lens must be selected to optimize the performance. The maximum NA of standard air objectives is typically found around 0.9, and the higher NA can only be achieved with liquid immersion objectives. For these lenses, immersion fluid with a refractive index greater than that of air ($n = 1$) is placed between the front lens and a coverslip. Common immersion media include water ($n \approx 1.33$) and oil ($n \approx 1.5$, depending on the formulation). Water or silicon oil immersion objectives are especially useful in confocal Raman microscopy for studying live cells in aqueous environments, as they minimize spherical aberrations and maintain high light collection efficiency when a focus is placed deeper in a sample. Regular oil immersion objectives are suited for measuring thin samples as the signal collection is the highest close to lens. Another important consideration while choosing an optimal lens is the working distance, defined as the distance between the front lens element to the focal point. Objectives with a high NA generally have shorter working distances, which can limit their use for samples with uneven surfaces or those enclosed in thick containers. In such cases, long-working distance objectives may be an option, although they are typically more costly and their NA is below 1. The balance between the NA and working distance is thus crucial for optimizing performance in different Raman microscopy applications. Finally, it is also important to check the light transmission property at the wavelength range of excitation laser and Raman spectrum to be measured. This is especially crucial when one builds a setup in near infrared (NIR) wavelength region. Common objective lenses do not transmit light in NIR region (~20% transmittance), and the majority of laser power introduced into the objective lens will be converted into heat within the lens itself.

2.1.3 Filters

Optical filters generally fall into two main categories based on their operating principles: absorptive filters (such as colored glass) and interference filters, which use thin dielectric coatings to achieve precise wavelength selectivity. In Raman spectroscopy, interference filters are commonly used due to their sharp spectral cutoffs and high transmission efficiency.

Three types of filters are needed for Raman microspectroscopy. One is a laser line filter to ensure that the spectral profile of an excitation laser is clean and monochromatic. Even when a single-frequency laser is used, it is recommended to use a laser line filter. At each optics that laser beam interacts with before reaching a sample (*e.g.*, mirrors, lenses, etc.), there is a chance of generating unwanted light, such as Raman scattering, from these optics. Thus, it is best to place this filter as close to the sample as possible. There is normally a dedicated place in a commercial filter cube of microscopes. Another is Raman edge long-pass filter-type dichroic mirror. Long-pass filters block light below the cutoff wavelength while transmitting the Raman signals (Stokes shifted photons) above the cutoff. This is placed at 45 degrees relative to the direction of a laser beam propagation. An important specification of dichroic mirror is its flatness. This is typically specified by fractions of wavelength of light. The flatness affects the quality of the beam profile (point spread function) when it is focused by objective lens. Typically, the Raman edge long-pass dichroic mirror blocks the excitation laser at OD6 ~ 8. Yet, another Raman edge long-pass filter is needed to block the excitation source completely.

Recently, a remarkable development was achieved in the field of optical filter technology. Ondax released a product line called SureBlock ultra-narrow band notch filters (the company later acquired by Coherent) [26]. Typical Notch filters can block a narrow band centered around the laser excitation wavelength with the bandpass of several nanometers. SureBlock filter can do it with 10 cm^{-1} bandwidth, enabling the acquisition of the low-frequency region of Raman spectrum, which contains rich information of phonon modes and intermolecular interactions [27, 28]. Furthermore, simultaneous detection of both Stokes and anti-Stokes Raman spectra at low-frequency region can be done, and it makes Raman thermometry easy. Switching a setup between normal Raman spectroscopy and low-frequency Raman spectroscopy can be easily done, thanks to the ultra-narrow notch filters. **Figure 2A** illustrates a scheme of a setup for low-frequency Raman measurement. The examples of spectra measured on the setup are shown in **Figure 2B–C**. **Figure 2B** shows the spectra of $SnCl_2$ aqueous solution at different laser power. The peak intensity of Stokes and anti-Stokes does not change as a function of laser power, which demonstrates that the temperature of water does not increase even at 1.2 W (https://doi.org/10.1073/pnas.2207713119). **Figure 2C** is the low-frequency Raman spectrum of D- and DL-alanine crystals. The difference between two crystals is clearer at low-frequency region (<300 cm^{-1}) than higher frequency (> 300 cm^{-1}) because phonon modes are highly sensitive to crystal structure.

2.1.4 Detectors

The most common way to measure Raman spectrum is the combined use of a spectrograph and a charge-coupled device (CCD). A commonly used spectrograph contains a slit, two spherical mirrors, and a diffraction grating. An important feature of a commercially available spectrograph is the focal length. For example, Andor sells spectrographs with focal lengths of 193 and 328 mm. The latter is more suited for Raman spectroscopy because the spectral resolution is higher. Then, the right selection of a diffractive grating is crucial to achieve the desired spectral resolution and bandwidth. When the spectral resolution is increased by using high-density gratings (*e.g.*, 2400 lines/mm), spectral range that can be measured by CCD becomes low. Thus, users need to design this aspect depending on experiments. Andor offers an online spectroscopy resolution and wavelength calculator (https://andor.oxinst.com/tools/resolution-calculator) to facilitate this task.

In most applications, a CCD based on silicon is used to capture signals spectrally resolved by spectrograph. Each pixel of a CCD generates an electronic charge upon absorption of photons. The greater the photon flux or exposure time, the more charge accumulates, resulting in a better signal-to-noise spectrum. The sensitivity of a CCD

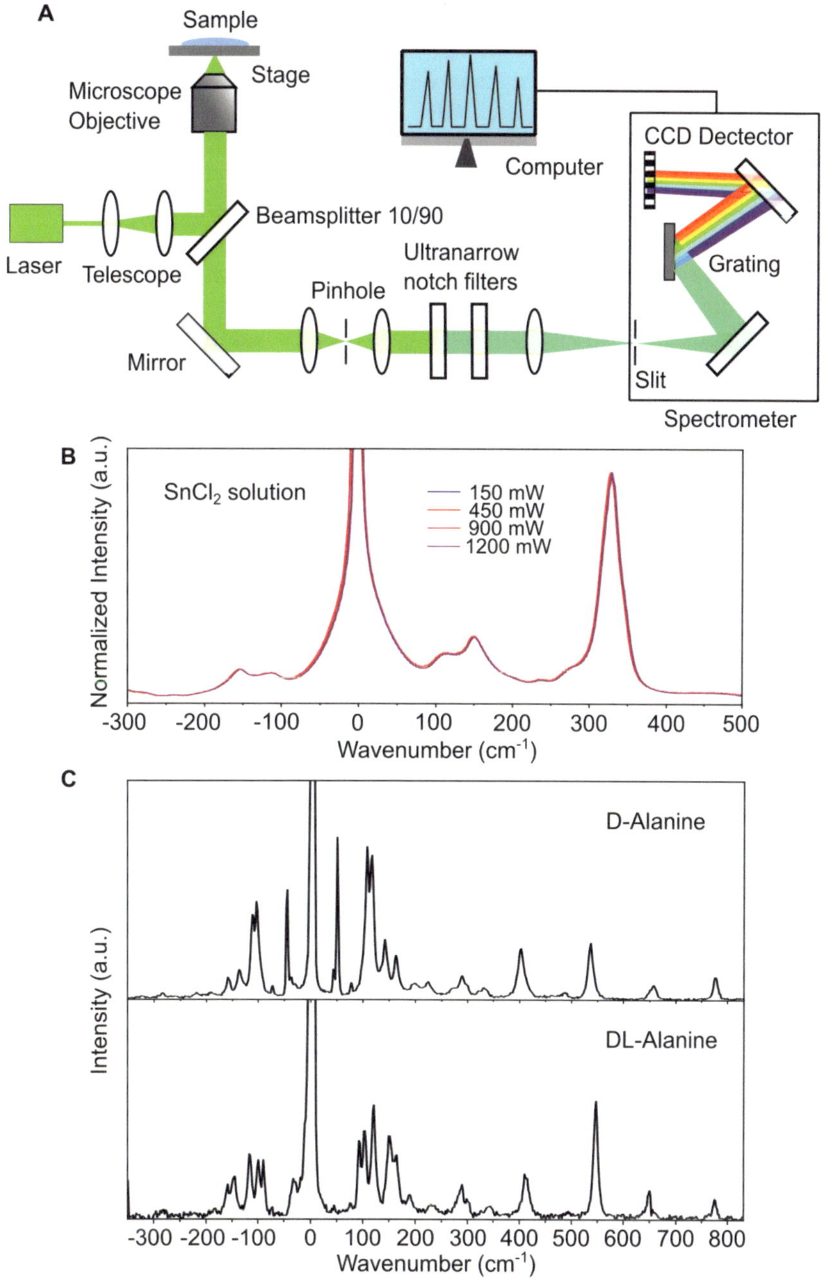

Figure 2.
(A) Schematic representation of a home-built setup for low-frequency Raman measurement. (B) Normalized low-frequency Raman spectra of $SnCl_2$ dissolved in water (1 mol·L^{-1}) measured at different laser powers. (C) Low-frequency Raman spectra of D- and DL-Alanine.

to incoming photons is defined by its quantum efficiency, making it a crucial parameter for evaluating detector performance. Dark current is also an important parameter to consider. Nowadays, CCDs are cooled by Peltier elements to minimize the dark noise. It can reach down to −100°C with the aid of water circulation (at 10°C). This is a huge cost and time saving as liquid nitrogen is not needed to cool CCDs.

For applications requiring enhanced sensitivity to reduce the data acquisition time, intensified charge-coupled devices (ICCDs) or electron-multiplied charge-coupled devices (EMCCDs) are employed. These detectors amplify the signal before readout, significantly improving the signal-to-noise ratio, especially in cases with very weak Raman scattering. One of the key parameters that affects the kinetics measurement (recording a series of spectra over the course of time) is the cycle time, which is the sum of the exposure time (the duration of signal collection) and the cleaning or readout time. Especially for time-resolved Raman spectroscopy application, it is ideal to have cycle time and exposure time to be identical, meaning no dead time of measurement due to cleaning period. Often, EMCCD has frame transfer function that achieves this goal. Additionally, gain settings are used to adjust the amplification of the signal: higher gain enhances sensitivity but can also increase noise (amplification noise such as EM gain noise). It is critical to balance these parameters depending on each application.

2.1.5 Microscope body

A microscope body serves as the central optical platform in a Raman microspectroscopy system, integrating an objective lens, a sample stage, a dichroic mirror cube, and a tube lens into a single stable structure. Both upright and inverted microscope configurations are used depending on the nature of the sample and experimental requirements. Upright microscopes, where an objective lens is positioned above the sample, are commonly employed for solid samples, thin films, and microstructured surfaces. In contrast, inverted microscopes, which have an objective positioned below the sample stage, offer distinct advantages for liquid-based applications. The choice of configuration is somewhat up to the preference of users. Inverted microscopes are often preferred in Raman tweezers setups involving aqueous samples or live cells [29]. It also simplifies the simultaneous use of high-NA immersion objectives and sample manipulation on the top-side, like adding reagents to a vial while observing. Thus, we often use an inverted microscope for *in situ* and real-time Raman studies of dynamic biological or crystallization processes.

2.1.6 Other components

In Raman microspectroscopy, several additional optical elements can be employed to optimize system performance and adapt the setup to specific experimental needs. Polarizers are sometimes used in both the excitation and detection paths to control the polarization state of light. Polarization-dependent Raman scattering can provide information regarding molecular orientation, anisotropy, and symmetry of systems. Telescopes, typically composed of two lenses arranged in a beam expander or reducer configuration, are used to adjust the diameter of a laser beam to overfill the back aperture of an objective lens. It is also important to perfectly collimate a beam to focus a beam at the right focal plane. Commercially available shearing interferometers (*e.g.*, sold by Thorlabs, Inc.) can be used to ensure the degree of collimation. Pinhole is a crucial element of confocal microscopy, acting as a spatial filter to remove the signal contribution from extra focal volume [23]. The size and position of a pinhole need to be carefully optimized. The size can be determined from the magnification of the objective lens used and the diameter of a focused beam at a sample.

A very important component that is not often discussed is the construction of a black box around a home-built setup. We have discussed how to optimize signal collection efficiency to maximize the signal-to-noise and signal-to-background ratio. If any stray light from a room light or displays of computers reaches detectors, the efforts are ruined as it affects background. In spectroscopy lab, it is a common practice to run measurements in dark, although this should not be a necessity. We still run measurements in relatively dark environments because we like to be in the dark. However, the measurement can be run without any problem even if our room lights are fully on. This is because we build a black box around detector area, and then the entire microscope setup is in a black box as well (**Figure 3**). The doors of the box are closed before every measurement so that no room light will be detected by our sensitive detectors. We measure a spectrum when a laser is not irradiated on a sample, which is so-called “dark” spectrum. This can be subtracted from spectra measured with a laser. This dark spectrum should be basically the same as the dark counts of CCD camera given by the specification sheet of the product. If it is not the case, there is some signal contamination due to stray lights. The construction of a good box is as important as all the other optics discussed above.

2.2 Calibration and spectral correction

Accurate calibration and spectral correction are essential for ensuring the reliability and reproducibility of Raman microspectroscopy measurements. Calibration should be performed often (daily) and especially when a major realignment of a setup is performed. Wavenumber calibration is typically performed using standard reference materials with well-characterized peaks, such as a mercury-neon lamp and a crystalline silicon (100) wafer [30]. Commercial spectrographs are equipped with

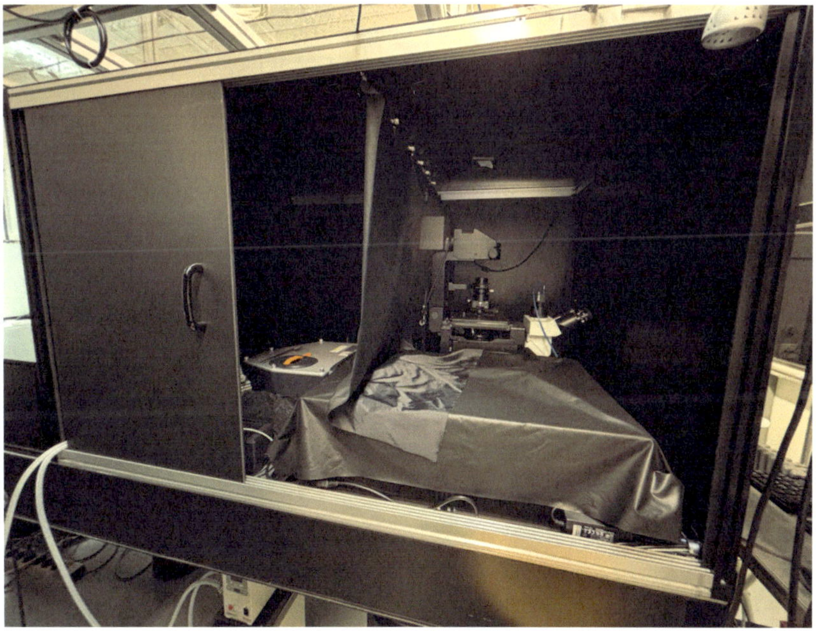

Figure 3.
Photo of a setup (“La neige”) in our group. It is placed in a black box to minimize interference from ambient light. Detectors are in another box (and black clothes to assist) to protect them from the stray light of an excitation laser in the box. This ensures only the Raman scattering from the sample is collected and the maximum signal-to-background ratio is achieved.

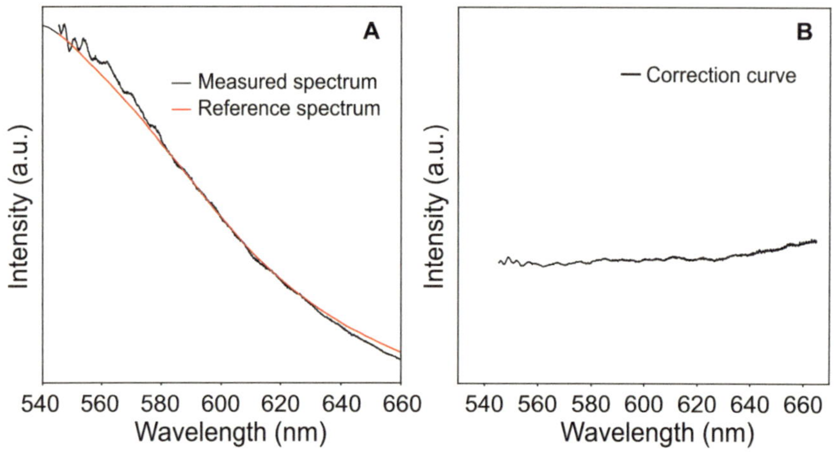

Figure 4.
(A) Fluorescence spectra of Coumarin 153 (C153) in methanol measured on our setup using a 532 nm laser. The measured fluorescence spectrum (black) and the reference data (red). (B) Spectral correction curve calculated from the data in the panel A. The nonuniform feature of the correction coefficients across the wavelength range is due to the wavelength-dependent sensitivity of the setup.

calibration function based on a single Raman peak from a silicon wafer, for example. We also perform a manual calibration using three Raman peaks from α-glycine to calibrate the spectral x-axis.

Besides wavelength calibration, intensity calibration should be performed as well. This is because the sensitivity of photon detection is not uniform over the wavelength range of interest. The major contribution to the wavelength-dependent variance is the quantum efficiency of CCD that is wavelength-dependent. Spectral profile of optical filters also contributes. Furthermore, the wavelength-dependent reflectivity and transmissivity of each optics (mirrors, lenses, polarizers, etc.) accumulate. The raw spectrum measured on CCD is the result of convolution of Raman spectrum and wavelength-dependent sensitivity of the setup. The calibration can be done by a calibrated white light source or intensity standard. In our group, we measure the fluorescence spectrum of Lamdachrome Dye series. Coumarin 153 (C153) in methanol is used for the intensity calibration of Raman spectrum measured using 532 nm laser [31]. For 405 and 785 nm, BBOT (LC4300) and Styryl 9 M (LC8400) are used, respectively. As shown in **Figure 4A**, the measured C153 spectrum does not match perfectly the reference data from literature. To compensate for the difference, we generate a spectral correction curve (**Figure 4B**), and this will be multiplied to a measured Raman spectrum during analysis.

3. Applications

A major advantage of a home-built Raman microspectroscopy setup is its tunability to adapt to various experimental demands without buying a commercial confocal Raman microscope one after another. For example, if an experiment were not compatible with the wavelength of a laser implemented in a setup, one can simply add another laser to the system with a proper set of optical filters. A standard commercial setup is often equipped with a single-frequency laser diode with a limited power output (50 ~ 100 mW). If one needs higher power for an experiment, it cannot be performed on a standard commercial microscope. In this section, we will introduce four different applications performed on the same microspectroscopy setup. For each application, different laser wavelength or power had to be used and therefore the setup was adapted accordingly.

3.1 Single crystal nucleation spectroscopy

Understanding the mechanism of crystal nucleation and polymorphism is a long-lasting problem yet to be solved. Rational control of crystallization is important in many industrial processes such as chemical, pharmaceutical, and food industries [32, 33]. A major challenge in studying crystal nucleation is its stochastic nature. It is impossible to predict when and where a nucleation event occurs, and therefore, one cannot capture the moment to observe its molecular dynamics. Recently, based on a Raman microspectroscopy setup, a new experimental approach was introduced to the field of the study of crystal nucleation. The so-called Single Crystal Nucleation Spectroscopy (SCNS) is a Raman microspectroscopy method applied to a solution of molecules at room temperature using a relatively high laser power (1.2 W) [22]. When a laser beam is focused in solution by a high-NA objective lens, particles feel optical gradient forces and are attracted toward the laser focus spot. For a microscale object, milliwatt laser power is enough to optically trap it stably. For small molecules such as amino acids, it was discovered that ~1 W focused laser beam can increase their concentration locally and a crystal can be formed at the laser spot (Optical Trapping Induced Crystallization: OTIC) [34]. While many OTIC works are performed by focusing a NIR laser, a 532 nm laser can also be used to confine a nucleation event under the laser focus. The benefit of using a 532 nm laser is that temperature increase of aqueous solution at laser spot is minute (~20 mK) and it generates Raman scattering simultaneously. Thus, *in situ* Raman spectral measurement can be realized during a single crystal nucleation event (**Figure 5A**).

Figure 5B–G shows the snapshots of Raman spectrum and optical microscopy image of a representative example during an SCNS experiment of glycine in water. In this experiment, time resolution of Raman spectroscopy was 46 ms (33 ms exposure time and 13 ms cleaning period) [22]. Initially, Raman spectrum shows the solution phase of glycine in water at the initial concentration of the sample prepared (**Figure 5B**). Near a nucleation event, Raman spectrum starts to change, which is seen as the intensity increase of glycine peaks and the decrease of the broad water bands (3100–3800 cm^{-1}). Upon nucleation, sharp Raman peaks appear and the observed new spectrum corresponds to the spectrum of β-glycine. It was widely accepted that glycine crystallization occurs via classical nucleation pathway where glycine forms cyclic dimers in solution and builds α-glycine crystal directly [35]. SCNS results revealed that glycine first forms metastable β-glycine via a nonclassical pathway involving prenucleation aggregates, and then α-glycine appears. In this study, non-negative matrix factorization (NMF) was used to analyze spectral evolution to extract a set of basis spectra during nucleation and the kinetics of each amplitude. High temporal resolution of Raman spectroscopy was necessary for the successful outcome of the NMF analysis and identifying the spectrum of prenucleation aggregates. This study highlights the power of time-resolved Raman microspectroscopy in revealing transient species during crystal formation. In a more recent study based on SCNS, NaCl was shown to significantly stabilize β-glycine, delaying its conversion to α-glycine and enabling secondary nucleation of γ-glycine on its surface [36]. The time resolution of Raman spectral acquisition was also improved to 20 ms per frame in this study due to the use of an EMCCD. The new knowledge obtained at the microscopic

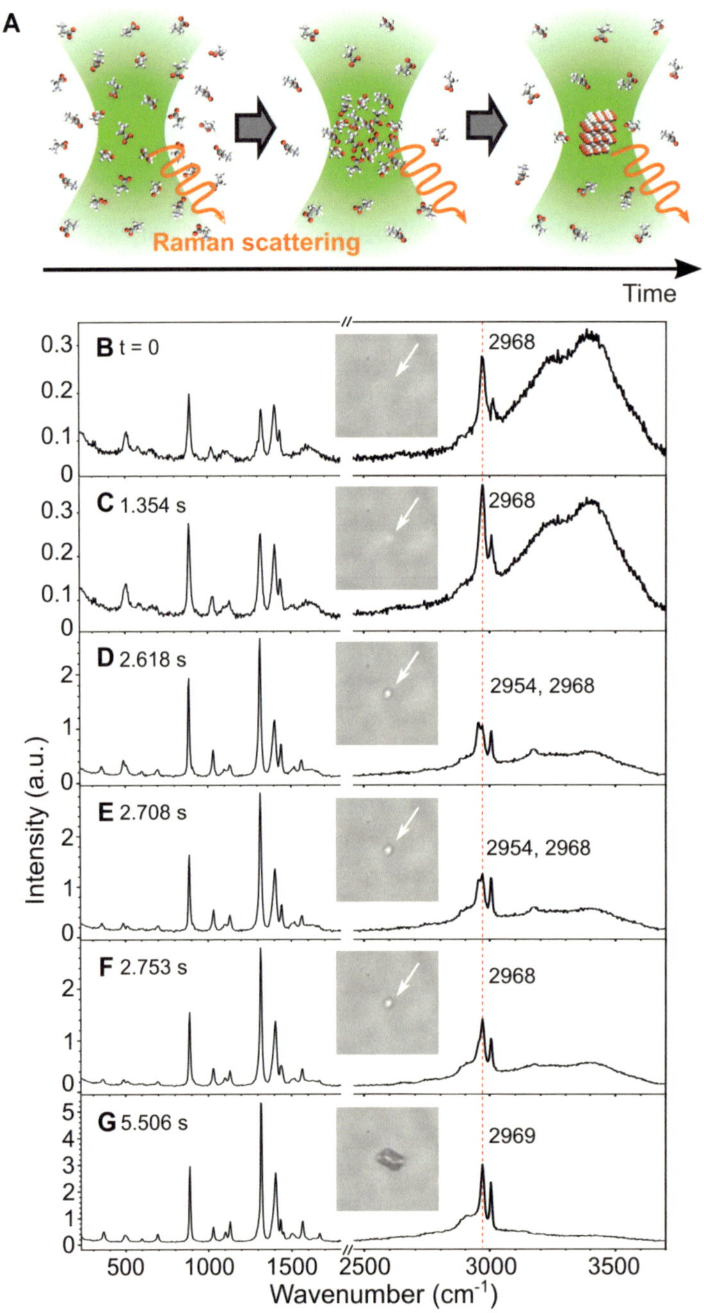

Figure 5.

A representative example of glycine nucleation in water at room temperature during an SCNS experiment. (A) A schematic illustration of the SCNS concept. (B-G) A series of time-resolved Raman spectra and bright field microscope images captured during a single glycine nucleation event in water. t = 0 s was arbitrarily defined near the nucleation point, which occurred approximately 2.5 minutes from the beginning of laser irradiation for this specific data set. In this example, time resolution of Raman spectroscopy was 46 ms (33 ms exposure time and 13 ms cleaning period) [22].

level will lead the field to achieve rational design and control of crystal nucleation and polymorphism, with applications ranging from pharmaceuticals to materials science.

3.2 Raman tweezer microspectroscopy for *in vivo* study of biomineralization

Raman tweezer microspectroscopy is a technique that has been used to characterize nano–/microparticles in air and solution [37–39]. It combines optical trapping

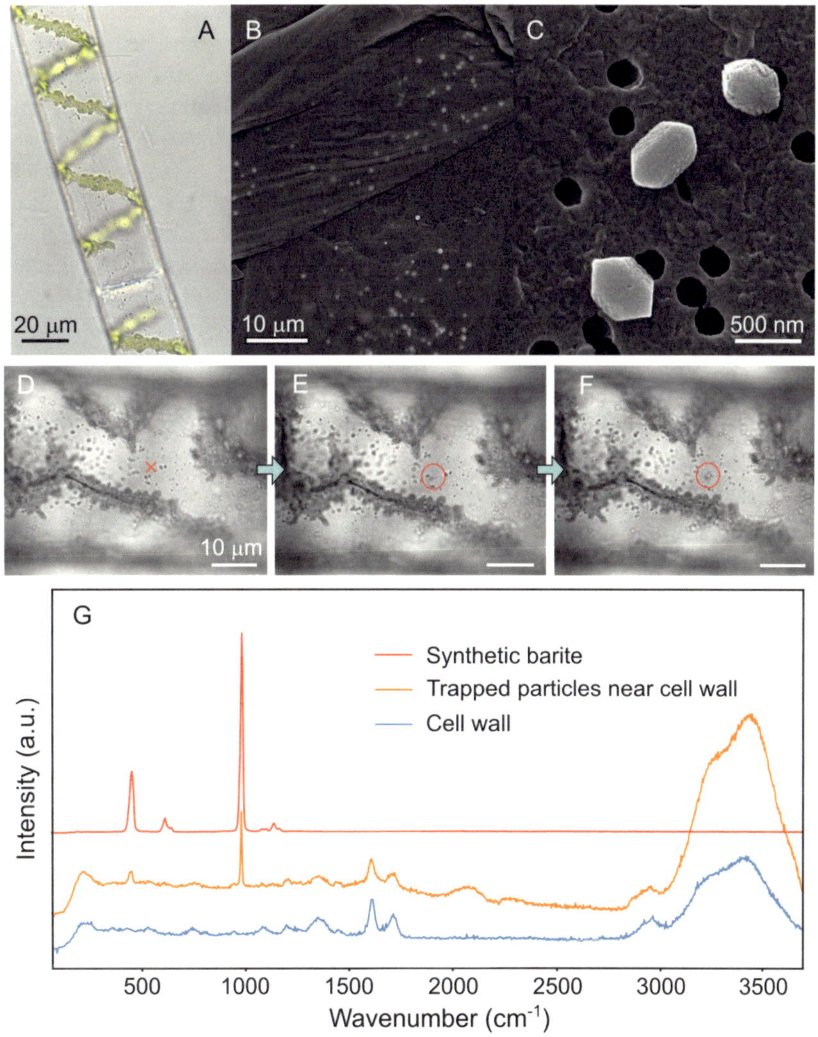

Figure 6.
(A) A bright field micrograph of Spirogyra E092. *(B-C) SEM images of barite microcrystals found in dried* Spirogyra *samples. (D-F) Successive optical microscopy images of living* Spirogyra *filaments showing the trapping of diffusive particles in the cells using a Raman tweezer microspectroscopy. The time interval is 0.5 seconds between each frame. The laser focal point prior to trapping is indicated by a red cross. When the laser was turned on, t particles move toward the laser focus, eventually forming an aggregate visible as a larger white particle trapped at the focal point. (G) Raman spectra collected from synthetic barite, trapped particles near a cell wall, and a cell wall. The trapped particles exhibit distinct Raman peaks that correspond closely to those of synthetic barite [40].*

and Raman spectroscopy as a powerful tool for real-time chemical analysis of trapped particles. Using a home-built Raman microspectroscopy setup, we studied barite biomineralization in freshwater alga, *Spirogyra* sp. [40]. In this study, the technique was employed to identify the location of barite ($BaSO_4$) microcrystals *in vivo* in *Spirogyra* (**Figure 6A**). It is not well understood why these microalgae form barite within the cells, and therefore, there are many fundamentally interesting questions regarding the formation mechanism and their functions. Scanning electron microscopy (SEM) images and Energy dispersive X-ray spectroscopy (EDX) measured from dried Spirogyra showed clear presence of barite microcrystals inside the cells (**Figure 6B–C**). However, the location of these crystals, whether they are in cytoplasm or vacuole, could not be determined from the measurements on dried samples. From the bright field microscopy observation of the cells of Spirogyra, we noticed the presence of many diffusive particles. In some literature, these particles were assigned as small refractive vesicles [41]. Initially, we regarded them as a type of vesicles and searched for barite near cell walls using Raman microspectroscopy. No Raman spectrum of barite was found after thorough investigations, and we decided to characterize the particles thought as vesicles using Raman spectroscopy. When a 532 nm CW laser (20–30 mW) was focused within the cells, these diffusive particles were optically trapped and accumulated at the focus spot (**Figure 6D–F**). Remarkably, the Raman spectrum measured from them matched with the barite Raman spectrum (**Figure 6G**), which enabled us to reassign them as barite, not vesicles. Having verified the intracellular location of barite crystals *in vivo*, we could provide insights into biomineralization mechanisms. As demonstrated here, laser power required to trap submicron-sized particles can be on the order of mW to tens of mW. At this range of power, spectra need to be accumulated over a few minutes to obtain a good signal-to-noise spectrum. Still, Raman tweezer microspectroscopy is a promising *in vivo* tool to obtain chemical information within biological systems. Raman microspectroscopy is emerging as a powerful *in vivo* tool for probing biomineralization processes [42].

3.3 *In situ* Raman microspectroscopy on calcium carbonate precipitation

Calcium carbonate is one of the most abundant materials on earth. Due to its various fascinating aspects related to how this world is shaped, it is studied in a wide range of disciplines such as chemistry, biology, geology, and material sciences. One of the intriguing questions at the molecular level is the complex precipitation mechanism involving a variety of crystal polymorphs [43]. The most stable polymorph is calcite and others are vaterite and aragonite, as well as amorphous solid so-called amorphous calcium carbonate (ACC) [43, 44]. In biology, one can find a combination of various polymorphs or even the metastable ACC used for different functions [45–47]. Despite the extensive studies, many questions remain on the formation process of calcium carbonate. The major difficulty arises from the strong correlation between the concentration of carbonate species and pH of aqueous solution. The composition of calcium carbonate species dynamically changes over time as precipitation proceeds and pH varies. *In situ* analysis of the precipitation process is a promising approach to follow chemical and structural information during precipitation.

Raman microspectroscopy can be used to investigate the complex pathways of calcium carbonate precipitation *in situ* [48]. On a Raman microspectroscopy setup built based on an inverted microscope body, the process of precipitation can be recorded by bright field microscopy techniques from the beginning of mixing calcium chloride and sodium carbonate. We can then characterize the sample any time we need *in situ* by Raman microspectroscopy. In this study, we screened a wide range of initial concentrations and pH of reactants and constructed a precipitation diagram [48]. We observed the formation of gel, ACC, and crystalline phases at different conditions, and followed how each system evolves over time. A variety of crystal morphologies was observed and it highlights the advantage of Raman spectroscopy to obtain their structural information *in situ* (**Figure 7**), as similar morphologies can be formed from different polymorphs (*e.g.*, dendritic vaterite vs. aragonite). Key advantages included the ability to monitor dynamic processes, such as gel-to-crystal transitions and Ostwald ripening, without disrupting the system. The ability to perform non-invasive, real-time chemical analysis at the microscale makes *in situ* Raman microspectroscopy an invaluable tool for studying complex processes such as crystallization pathways, polymorphic transitions, and reaction kinetics.

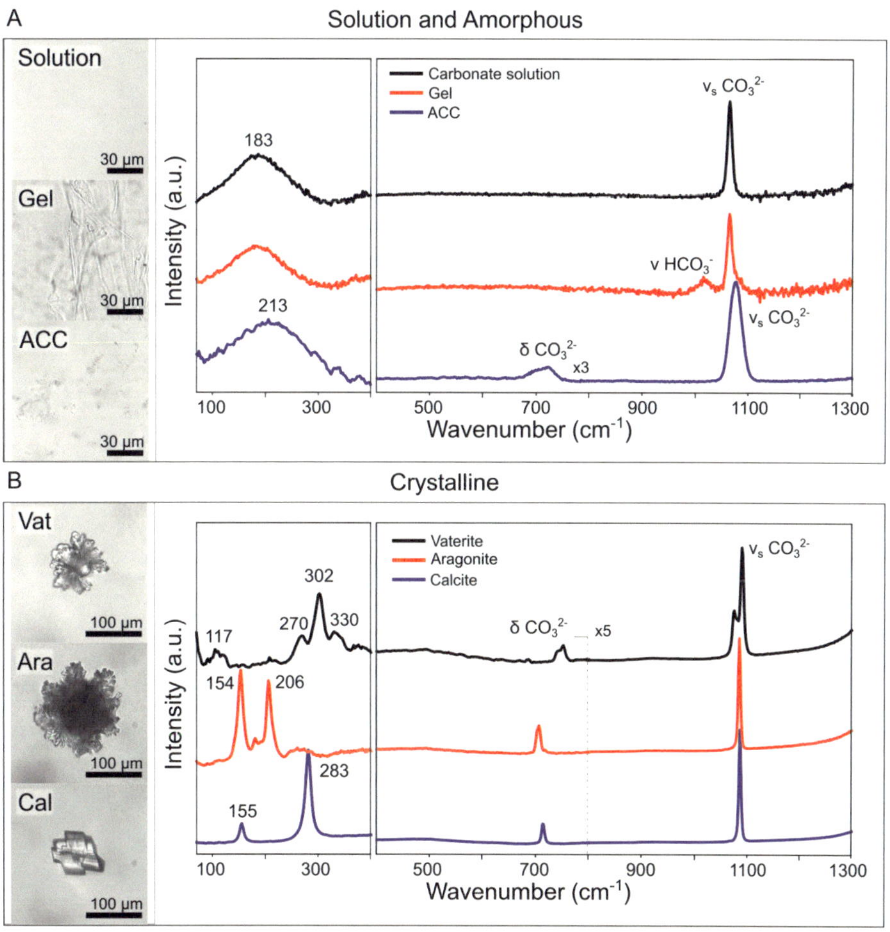

Figure 7.
Characterization of amorphous and crystalline phases of calcium carbonate using Raman microspectroscopy. (A) Raman spectrum of calcium carbonate, (top) solution phase, (middle) gel, and (bottom) amorphous calcium carbonate (ACC). Optical images show the corresponding area of the measurement. (B) Raman spectrum of the three crystalline polymorphs. Optical images show an example of each polymorph, while the spectrum shown on the right corresponds to the exact crystal on the left. The intensity of 400–800 cm^{-1} spectral region was enhanced fivefold to improve the peak visibility [48].

3.4 Raman microspectroscopy for environmental research

The detection of micro−/nano-plastics and understanding the degree of pollution in the environment have become urgent tasks recently. The severe state of pollution in the environment and human bodies has been reported and discussed as an emergent threat to human society. Raman spectroscopy has been used for the characterization of chemical composition of plastics at macro, micro, and nanoscale [49, 50]. Recently, not only the composition of plastics but also the presence of inorganic compounds within plastics became aware in the community [51, 52]. Titanium oxide is an example widely found in many plastic products. While it has been regarded as a nontoxic substance and used in many products as pigments, flame retardants, and food additives, recently the European Union changed its view and banned its use as food additives

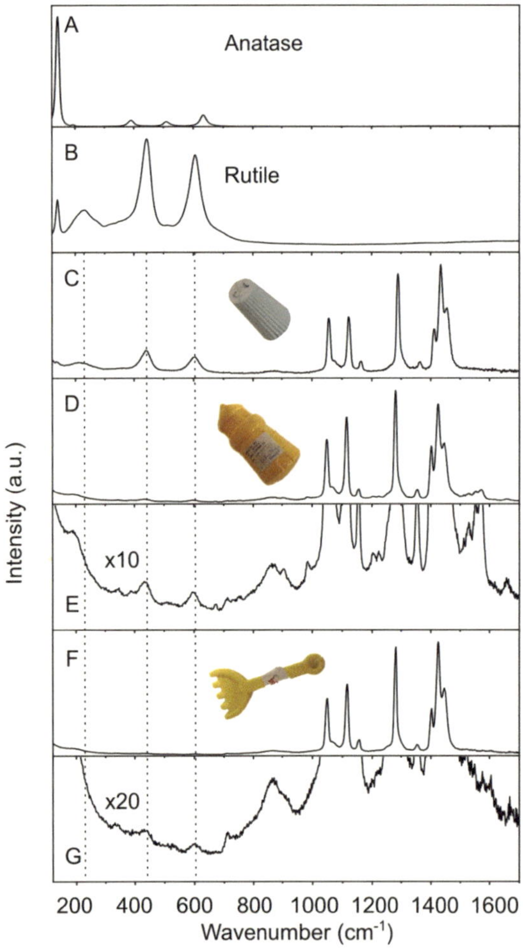

Figure 8.
Detection of TiO_2 in HDPE plastic materials with varying concentrations of elemental titanium using Raman microspectroscopy (785 nm laser excitation). (A-B) Raman spectrum of reference anatase and rutile TiO_2 for comparison. (C) Spectrum of HDPE sample containing 8661 mg·kg^{-1} Ti. (D–E) Spectrum of sample with 1145 mg·kg^{-1} Ti, where the panel E shows a 10× magnified view of D. (F–G) Spectrum of sample with 445 mg·kg^{-1} Ti, with the panel G being a 20× magnified version of F. The spectra demonstrate that characteristic Raman bands of rutile TiO_2 can still be detected at lower Ti concentrations when appropriately scaled [52].

already. We demonstrated Raman microspectroscopy as a sensitive tool to identify the polymorphs of titanium oxide embedded in various plastic consumer goods [51, 52].

Figure 8 shows Raman spectra of three high-density polyethylene (HDPE) samples containing varying concentrations of titanium, ranging from approximately 8660 to 450 ppm (mg·kg^{-1}) as determined by X-ray fluorescence (XRF) [52]. Raman spectroscopy revealed that the titanium is present in the form of rutile-phase TiO_2. It is remarkable that the method could measure the spectrum of TiO_2 while its concentration can be as low as 450 ppm. Due to pigments used for coloring plastic consumer goods, some samples show strong fluorescence [53]. On a home-built setup, it is possible to add another wavelength of laser to circumvent this problem. Initially, we used 532 and 785 nm as an excitation laser source. For the samples which fluoresced on both wavelength, initially 830 and 900 nm were implemented with little success on eliminating fluorescence completely. Then, we implemented 405 nm laser as another source. The idea behind this choice is that some compounds start the on-set of their fluorescence spectrum above 430 nm, and it may be possible to record Raman spectrum between 405 and 430 nm (corresponding to spectral range up to 1400 cm^{-1}). In fact, the detection of TiO_2 and the assignment of polymorph were easier at 405 nm laser setup for some samples [52]. Alternatively, there is another approach that has been proposed as to use time-gated detection to discriminate Raman scattering from Fluorescence backgrounds [54]. These studies underscore the growing role of Raman microspectroscopy in environmental monitoring and pollution assessment, where detailed chemical identification at the microscale is essential.

4. Conclusion

In this chapter, we described the advantages of Raman microspectroscopy over conventional Raman spectroscopy for more efficient signal collection, higher spatial resolution, and better signal-to-background measurement. The proper design of a setup and assembly of a CW laser, an objective lens, pinhole, and filters, *etc.* can provide a powerful tool for scientists. A well-constructed Raman microspectroscopy setup allows rapid and high-quality spectral acquisition, which is essential for advanced applications. Through careful consideration of each component in the setup and taking advantage of the tunability of a home-built setup, users can tailor the system for various tasks without buying a new commercial setup. We highlighted several applications, such as monitoring phase transitions via time-resolved Raman measurements, determining the location of biominerals within cells *in vivo*, identifying crystal polymorphs *in situ*, and detecting nanocrystals embedded in complex matrices like plastics. While it may have seemed challenging to construct a home-built Raman microspectroscopy setup, we hope this chapter will change this impression of readers. Besides the huge reduction of cost to obtain a Raman microscope by building one than buying a commercial one, the tunability of the setup is priceless for scientists working at the frontier of creating knowledge. A home-built Raman microspectroscopy setup can be a powerful means to discover knowledge that a commercial one may not let you reach.

Acknowledgements

Z.L. and T.B.M.A. are thankful for the financial support from the SNSF (200021_219291) and the University of Geneva. Z.L. thanks Xi'an University of Architecture and Technology for the scholarship in the "Young Scholars Project".

Conflict of interest

The authors declare no conflict of interest.

Author details

Zihan Li and Takuji B.M. Adachi*
Department of Physical Chemistry, University of Geneva, Geneva, Switzerland

*Address all correspondence to: takuji.adachi@unige.ch

References

[1] Tittel J, Knechtel F, Ploetz E. Conquering metal–organic frameworks by Raman scattering techniques. Advanced Functional Materials. 2024;**34**:2307518. DOI: 10.1002/adfm.202307518

[2] Wei X, Jie D, Cuello JJ, Johnson DJ, Qiu Z, He Y. Microalgal detection by Raman microspectroscopy. TrAC Trends in Analytical Chemistry. 2014;**53**:33-40. DOI: 10.1016/j.trac.2013.09.012

[3] Kuhar N, Sil S, Umapathy S. Potential of Raman spectroscopic techniques to study proteins. Spectrochimica Acta Part A: Molecular and Biomolecular Spectroscopy. 2021;**258**:119712. DOI: 10.1016/j.saa.2021.119712

[4] Anger PM, Von Der Esch E, Baumann T, Elsner M, Niessner R, Ivleva NP. Raman microspectroscopy as a tool for microplastic particle analysis. TrAC Trends in Analytical Chemistry. 2018;**109**:214-226. DOI: 10.1016/j.trac.2018.10.010

[5] Rygula A, Majzner K, Marzec KM, Kaczor A, Pilarczyk M, Baranska M. Raman spectroscopy of proteins: A review. Journal of Raman Spectroscopy. 2013;**44**:1061-1076. DOI: 10.1002/jrs.4335

[6] Das RS, Agrawal YK. Raman spectroscopy: Recent advancements, techniques and applications. Vibrational Spectroscopy. 2011;**57**:163-176. DOI: 10.1016/j.vibspec.2011.08.003

[7] Chrimes AF, Khoshmanesh K, Stoddart PR, Mitchell A, Kalantar-zadeh K. Microfluidics and Raman microscopy: Current applications and future challenges. Chemical Society Reviews. 2013;**42**:5880. DOI: 10.1039/c3cs35515b

[8] Byrne HJ, Sockalingum GD, Stone N. Raman microscopy: Complement or competitor? In: Moss D, Barnett NW, editors. Biomedical Applications of Synchrotron Infrared Microspectroscopy. Cambridge, UK: The Royal Society of Chemistry; 2010. pp. 105-143. DOI: 10.1039/9781849731997-00105

[9] Huang WE, Li M, Jarvis RM, Goodacre R, Banwart SA. Shining light on the microbial world. Advances in Applied Microbiology. 2010;**70**:153-186. DOI: 10.1016/S0065-2164(10)70005-8

[10] Kano H, Segawa H, Leproux P, Couderc V. Linear and nonlinear Raman microspectroscopy: History, instrumentation, and applications. Optical Review. 2014;**21**:752-761. DOI: 10.1007/s10043-014-0123-9

[11] Peterson W, De Pablo JG, Lindley M, Hiramatsu K, Goda K. Ultrafast impulsive Raman spectroscopy across the terahertz–fingerprint region. Advanced Photonics. 2022;**4**:016003. DOI: 10.1117/1.AP.4.1.016003

[12] Finneran IA, Welsch R, Allodi MA, Miller TF, Blake GA. Coherent two-dimensional terahertz-terahertz-Raman spectroscopy. Proceedings of the National Academy of Sciences USA. 2016;**113**:6857-6861. DOI: 10.1073/pnas.1605631113

[13] Jing Y, Zhang J, Wan M, Xue J, Liu J, Qin J, et al. Distinguishing polymorphs of ethenzamide-saccharin cocrystal based on terahertz and Raman vibrational spectroscopic techniques. IEEE Transactions on Terahertz Science and Technology. 2024;**14**:152-161. DOI: 10.1109/TTHZ.2024.3358734

[14] Cheng J-X, Wei M, Ozeki Y, Polli D, editors. Stimulated Raman

Scattering Microscopy: Techniques and Applications. Amsterdam: Elsevier; 2021

[15] Cheng J-X, Xie XS. Coherent anti-stokes Raman scattering microscopy: Instrumentation, theory, and applications. The Journal of Physical Chemistry. B. 2004;**108**:827-840. DOI: 10.1021/jp035693v

[16] Xu P, Suntivich J. Time-resolved monitoring of electrochemical reactions using in situ stimulated Raman spectroscopy. ACS Sustainable Chemistry & Engineering. 2023;**11**:13-17. DOI: 10.1021/acssuschemeng.2c04873

[17] Gracin D, Štrukil V, Friščić T, Halasz I, Užarević K. Laboratory real-time and in situ monitoring of mechanochemical milling reactions by Raman spectroscopy. Angewandte Chemie, International Edition. 2014;**53**:6193-6197. DOI: 10.1002/anie.201402334

[18] Chang C, Feng L-F, Gu X-P, Zhang C-L, Dai L-K, Chen X, et al. In situ Raman spectroscopy real-time monitoring of a polyester polymerization process for subsequent process optimization and control. Industrial and Engineering Chemistry Research. 2022;**61**:17993-18003. DOI: 10.1021/acs.iecr.2c02933

[19] Liao C-S, Cheng J-X. In situ and in vivo molecular analysis by coherent Raman scattering microscopy. Annual Review of Analytical Chemistry. 2016;**9**:69-93. DOI: 10.1146/annurev-anchem-071015-041627

[20] Wang J, Alieva A, Boyes M, Pollard AJ, Casiraghi C. In situ Raman study of the crystallization of glycine. CrystEngComm. 2023;**25**:2591-2599. DOI: 10.1039/D3CE00241A

[21] Nanubolu JB, Burley JC. In situ Raman mapping for identifying transient solid forms. CrystEngComm. 2015;**17**:5280-5287. DOI: 10.1039/C5CE00008D

[22] Urquidi O, Brazard J, LeMessurier N, Simine L, Adachi TBM. In situ optical spectroscopy of crystallization: One crystal nucleation at a time. Proceedings of the National Academy of Sciences USA. 2022;**119**:e2122990119. DOI: 10.1073/pnas.2122990119

[23] Wilson T. Resolution and optical sectioning in the confocal microscope: Properties of the fluorescent confocal microscope. Journal of Microscopy. 2011;**244**:113-121. DOI: 10.1111/j.1365-2818.2011.03549.x

[24] Zhu X, Xu T, Lin Q, Duan Y. Technical development of Raman spectroscopy: From instrumental to advanced combined technologies. Applied Spectroscopy Reviews. 2014;**49**:64-82. DOI: 10.1080/05704928.2013.798801

[25] Hughes DW, Barr JRM. Laser diode pumped solid state lasers. Journal of Physics D: Applied Physics. 1992;**25**:563-586. DOI: 10.1088/0022-3727/25/4/001

[26] Moser C, Havermeyer F. Ultra-narrow-band tunable laserline notch filter. Applied Physics B: Lasers and Optics. 2009;**95**:597-601. DOI: 10.1007/s00340-009-3447-6

[27] Bensley KA, Chimenti RV, Lofland SE. Low-frequency Raman spectroscopy of polymers. Macro Chemistry & Physics. 2025;**226**:70004. DOI: 10.1002/macp.202400517

[28] Larkin PJ, Dabros M, Sarsfield B, Chan E, Carriere JT, Smith BC. Polymorph characterization of active pharmaceutical ingredients (APIs) using low-frequency Raman spectroscopy. Applied Spectroscopy. 2014;**68**:758-776. DOI: 10.1366/13-07329

[29] Casabella S, Scully P, Goddard N, Gardner P. Automated analysis of single cells using laser tweezers Raman spectroscopy. Analyst. 2016;**141**:689-696. DOI: 10.1039/C5AN01851J

[30] Itoh N, Shirono K. Reliable estimation of Raman shift and its uncertainty for a non-doped Si substrate (NMIJ CRM 5606-a). Journal of Raman Spectroscopy. 2020;**51**:2496-2504. DOI: 10.1002/jrs.6003

[31] Brackmann U. Lambdachrome Laser Dyes. 3rd ed. Göttingen, Germany: Lambda Physik; 1986

[32] Mullin JW. Crystallization. 4th ed. Oxford; Boston: Butterworth-Heinemann; 2001

[33] Bernstein J. Polymorphism in Molecular Crystals. 2nd ed. New York, USA: Oxford University Press; 2020. DOI: 10.1093/oso/9780199655441.001.0001

[34] Sugiyama T, Adachi T, Masuhara H. Crystallization of glycine by photon pressure of a focused CW laser beam. Chemistry Letters. 2007;**36**:1480-1481. DOI: 10.1246/cl.2007.1480

[35] Weissbuch I, Torbeev VY, Leiserowitz L, Lahav M. Solvent effect on crystal polymorphism: Why addition of methanol or ethanol to aqueous solutions induces the precipitation of the least stable β form of glycine. Angewandte Chemie, International Edition. 2005;**44**:3226-3229. DOI: 10.1002/anie.200500164

[36] Cotting G, Urquidi O, Besnard C, Brazard J, Adachi TBM. The effect of salt additives on the glycine crystallization pathway revealed by studying one crystal nucleation at a time. Proceedings of the National Academy of Sciences USA. 2025;**122**:e2419638122. DOI: 10.1073/pnas.2419638122

[37] Buajarern J, Mitchem L, Reid JP. Manipulation and characterization of aqueous sodium dodecyl sulfate/sodium chloride aerosol particles. The Journal of Physical Chemistry. A. 2007;**111**:13038-13045. DOI: 10.1021/jp0764034

[38] Chan JW. Recent advances in laser tweezers Raman spectroscopy (LTRS) for label-free analysis of single cells. Journal of Biophotonics. 2013;**6**:36-48. DOI: 10.1002/jbio.201200143

[39] Mrad R, Kruglik SG, Ben Brahim N, Ben Chaâbane R, Negrerie M. Raman tweezers microspectroscopy of functionalized 4.2 nm diameter CdSe nanocrystals in water reveals changed ligand vibrational modes by a metal cation. Journal of Physical Chemistry C. 2019;**123**:24912-24918. DOI: 10.1021/acs.jpcc.9b06756

[40] Barbosa N, Jaquet J-M, Urquidi O, Adachi TBM, Filella M. Combined in vitro and in vivo investigation of barite microcrystals in Spirogyra (Zygnematophyceae, Charophyta). Journal of Plant Physiology. 2022;**276**:153769. DOI: 10.1016/j.jplph.2022.153769

[41] Grolig F. Actin-based organelle movements in interphase *Spirogyra*. Protoplasma. 1990;**155**:29-42. DOI: 10.1007/bf01322613

[42] Pilátová J, Pánek T, Oborník M, Čepička I, Mojzeš P. Revisiting biocrystallization: Purine crystalline inclusions are widespread in eukaryotes. The ISME Journal. 2022;**16**:2290-2294. DOI: 10.1038/s41396-022-01264-1

[43] Du H, Amstad E. Water: How does it influence the $CaCO_3$ formation? Angewandte Chemie, International Edition. 2020;**59**:1798-1816. DOI: 10.1002/anie.201903662

[44] Nielsen MH, Aloni S, De Yoreo JJ. In situ TEM imaging of $CaCO_3$ nucleation reveals coexistence of direct and indirect pathways. Science. 2014;**345**:1158-1162. DOI: 10.1126/science.1254051

[45] Sommerdijk NAJM, With GD. Biomimetic $CaCO_3$ mineralization using designer molecules and interfaces. Chemical Reviews. 2008;**108**:4499-4550. DOI: 10.1021/cr078259o

[46] Dey A, De With G, Sommerdijk NAJM. In situ techniques in biomimetic mineralization studies of calcium carbonate. Chemical Society Reviews. 2010;**39**:397-409. DOI: 10.1039/b811842f

[47] Gilbert PUPA, Bergmann KD, Boekelheide N, Tambutté S, Mass T, Marin F, et al. Biomineralization: Integrating mechanism and evolutionary history. Science Advances. 2022;**8**:eabl9653. DOI: 10.1126/sciadv.abl9653

[48] Barbosa N, Khakhula T, Brazard J, Adachi TBM. Beauty and complexity of calcium carbonate precipitation: Optical microscopy and in situ Raman microspectroscopy characterization. Crystal Growth & Design. 2025;**25**:210-224. DOI: 10.1021/acs.cgd.4c01063

[49] Miller EA, Yamahara KM, French C, Spingarn N, Birch JM, Van Houtan KS. A Raman spectral reference library of potential anthropogenic and biological ocean polymers. Scientific Data. 2022;**9**:780. DOI: 10.1038/s41597-022-01883-5

[50] Ribeiro-Claro P, Nolasco MM, Araújo C. Characterization of microplastics by Raman spectroscopy. Comprehensive Analytical Chemistry. 2017;**75**:119-151. DOI: 10.1016/bs.coac.2016.10.001

[51] Filella M, Brazard J, Adachi TBM, Turner A. Hazardous chemical elements in cleaning cloths, a potential source of microfibres. Science of the Total Environment. 2022;**846**:157419. DOI: 10.1016/j.scitotenv.2022.157419

[52] Brazard J, Adachi TBM, Turner A, Filella M. Determination of titanium speciation in consumer plastics by Raman microspectroscopy. Microchemical Journal. 2025;**208**:112391. DOI: 10.1016/j.microc.2024.112391

[53] Azari A, Ronsmans S, Vanoirbeek JAJ, Hoet PHM, Ghosh M. Challenges in Raman spectroscopy of (micro)plastics: The interfering role of colourants. Environmental Pollution. 2024;**363**:125250. DOI: 10.1016/j.envpol.2024.125250

[54] Kögler M, Heilala B. Time-gated Raman spectroscopy – A review. Measurement Science and Technology. 2020;**32**:012002. DOI: 10.1088/1361-6501/abb044

Chapter 3

Terahertz Raman Spectroscopy of Ferroelectric Materials

Seiji Kojima

Abstract

Ferroelectric materials have been used to many kinds of devices by rich of functionalities such as pyroelectric, optical, photocatalytic, and piezoelectric effects. Ferroelectricity is induced by the anharmonic lattice distortion. Soft optic modes of displacive-type ferroelectrics and fast relaxation processes (central peaks) of order-disorder-type ferroelectrics have been investigated by terahertz Raman spectroscopy ($3 \sim 200\ cm^{-1}$), which is comparable with the current terahertz time-domain spectroscopy ($0.1 \sim 6$ THz). This chapter reviews the terahertz Raman spectroscopy of the technologically important ferroelectric materials. This chapter reviews the temperature dependences of ferroelectric soft modes and central peaks in a large temperature range, and the dispersion of phonon-polariton of ferroelectric crystals with the perovskite and related structures.

Keywords: Raman spectroscopy, terahertz, ferroelectrics, soft mode, central peak, polariton, phase transition

1. Introduction

1.1 Terahertz dynamics on ferroelectric materials

When a strain appears by an external electric field on a crystal and an electric polarization appears by an external stress field, this phenomenon is known as the piezoelectric effect. Piezoelectricity was discovered in α-quartz and Rochelle salt (sodium potassium tartrate tetrahydrate) crystals in 1880 [1]. When a spontaneous polarization exists without an external electric field and its reversal occurs by an external electric field, it is called ferroelectricity. In 1921, the experimental evidence of ferroelectricity was reported for the first time on Rochelle salt [2]. Ferroelectric materials are known for various kinds of applications such as nonvolatile memory, piezoelectric, electrostrictive, pyroelectric, acousto-optic, electro-optic, photorefractive, and nonlinear optic devices.

Ferroelectric materials are insulators with a high energy gap, and the lattice dynamics plays a dominant role. Ferroelectric instability is divided into two kinds of crystal lattice instability. In displacive type ferroelectrics, the freezing of atomic displacements induces a spontaneous polarization, and the mode frequency of the corresponding optic mode softens toward the ferroelectric Curie temperature, T_C. In order-disorder type ferroelectrics, a spontaneous polarization appears by the ordering of dipoles below T_C. The relaxation time of polarization fluctuations diverges toward T_C, and it is called a critical slowing down. The polarization fluctuations are observed as fast relaxation processes in a Raman scattering spectrum and are called a central peak (or quasi-elastic scattering). These dynamical properties of soft modes and central peaks are related to the ferroelectric instability observed in the THz range. Their temperature and pressure dependences of THz vibrational properties have been studied by Raman scattering, far-infrared spectroscopy, and terahertz time-domain (THz-TDS) spectroscopy [3, 4].

1.2 THz Raman spectroscopy

Brillouin predicted the inelastic light by density fluctuations associated with sound waves in a fluid in 1922. Brillouin scattering is known as the inelastic scattering by acoustic phonons. The frequency spectrum consists of a doublet split in a gigahertz range around the incident light, and sound velocity is determined by its frequency shift [5, 6]. The light scattering by optical vibrational modes in liquids such as benzene was reported by Raman and Krishnan in 1928 [7]. Raman scattering is known as the inelastic light scattering by optical phonons. Both frequency spectra include a central unshifted component due to the nonpropagating entropy fluctuations. It is referred to as Rayleigh scattering. **Figure 1** shows the process of Raman scattering. Incident light to a sample is monochromatic with the frequency ν_i and wavevector $\vec{k}_i$. It is inelastically scattered by thermally excited optical phonons with the frequency ν_R and wavevector $\vec{q}$ in a sample. According to the energy and momentum conservation, it holds that for the Stokes component,

$$\nu_R = \nu_i - \nu_s, \tag{1}$$

$$\vec{q} = \vec{k}_i - \vec{k}_s, \tag{2}$$

where ν_s and $\vec{k}_s$ are the frequency and wavevector of scattered light, respectively.

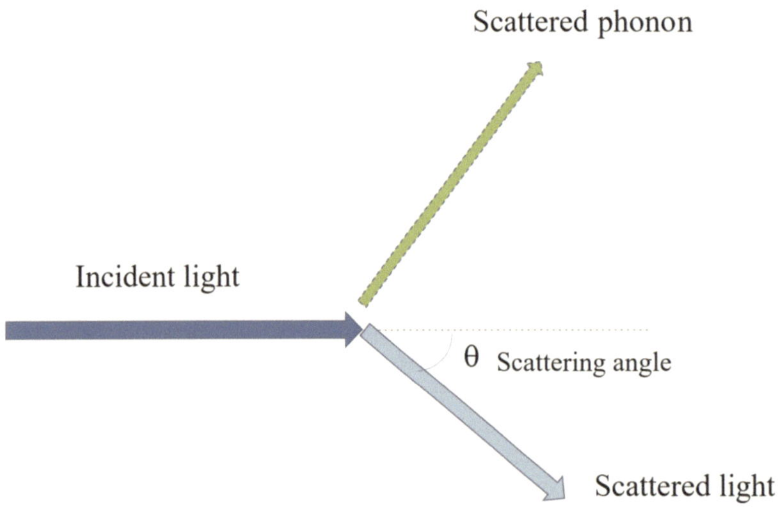

Figure 1.
Process of Raman scattering. of wavevectors Incident light is scattered by thermally excited phonons in a sample. Frequency spectra depend on scattering angle.

After the development of a laser in the 1960s, laser-Raman scattering using a grating spectrometer and a photon-counting system has been applied in various fields. In conventional Raman equipment, the observable frequency is usually from 100 to 4000 cm^{-1}, and the measurement below 100 cm^{-1} is difficult due to the unshifted intense elastic scattering. However, it is necessary to measure terahertz Raman spectra lower than 100 cm^{-1} for the observation of the soft modes and central peaks in ferroelectric materials. For the low-frequency measurement, the spatial techniques such as vapor filters, spatial filters, and a triple-grating spectrometer were used to reduce the elastic scattering. Recently, super-notch filters [8] have been developed, and the low-frequency measurement is possible using a single-grating spectrometer down to 3 cm^{-1}. Nowadays, the low-frequency Raman scattering below 200 cm^{-1} is called "THz Raman scattering" [9, 10].

2. Perovskite oxides

In 1942, the dielectric anomaly related to ferroelectricity was discovered in chemically stable $BaTiO_3$ with perovskite structure, which is originated by mineral perovskite ($CaTiO_3$) [3]. Its piezoelectricity has attracted much attention. The great advantage is the rich of substitution in A and B sites of the perovskite structure, where A is a cation located in the outside of an oxygen octahedron, and B is a cation located in an oxygen octahedron. The various ferroelectric and antiferroelectric materials have been discovered in $A^{1+}B^{5+}O^{2-}{}_3$ and $A'^{2+}B'^{4+}O^{2-}{}_3$ compounds such as $PbTiO_3$, $PbZrO_3$, $NaNbO_3$, $KNbO_3$. The structural phase transitions of these perovskite ferroelectric oxides are caused by a lattice dynamical instability [10].

In this section, as examples of displacive-type phase transitions, soft optic modes were described in $LaAlO_3$ and $SrHfO_3$. As an example of the ferroelectric phase transition of the order-disorder type, the critical slowing down of $(Na_{0.5}Bi_{0.5})NbO_3$ is described.

2.1 $LaAlO_3$

Recently, vacuum-ultraviolet photodetectors have attracted much attention by the application to space exploration and semiconductor devices. Benefiting from its intrinsic ultra-wide band-gap, chemical robustness, and low-cost features, $LaAlO_3$ shows great promise in developing next-generation compact, cheap, and easy-to-fabricate VUV PDs [11]. The rhombohedral $LaAlO_3$ with perovskite structure has the symmetry of the space group $R\bar{3}c$ (Z = 2), where Z is the number of formula units per primitive cell [12]. $LaAlO_3$ undergoes an improper ferroelastic phase transition from the cubic phase ($Pm\bar{3}m$, Z = 1) to ferroelastic nonpolar rhombohedral phase ($R3c$, Z = 2) at T_C = 813 K [13]. The rhombohedral structure appears from the cubic structure by rotation of the adjacent AlO_6 octahedra in opposite directions around the $[111]_c$ cubic direction, and it belongs to the family of rotationally distorted perovskites with Glazer's notation ($a^-a^-a^-$) [14]. In the cubic-rhombohedral phase transition, the order parameter is the tilt angle of an AlO_6 oxygen octahedron along the [111] direction.

In a cubic phase, the soft mode is the triply degenerated R_{2u} mode at an R point. In the rhombohedral phase, the soft modes are nondegenerated A_{1g} and doubly degenerated E_g modes at an R point [15]. They are Raman-active, while infrared inactive modes. The displacement of A_{1g} and E_g modes are along the [111] direction and in the plane perpendicular to the [111] axis, respectively. The Raman scattering of A_{1g} and E_g modes was observed as a function of temperature. The low-frequency spectra of E_g modes of a $LaAlO_3$ crystal observed at backscattering geometry are shown in **Figure 2**. **Figure 3** shows the squared mode frequency of A_{1g} and E_g modes as a function of temperature observed by Raman scattering. These modes show the remarkable softening toward T_C = 813 K [16]. This fact indicates a highly displacive nature for the phase transition at T_C. Some anomaly was reported at about T_d = 730 K, and above T_d, the dynamic disorder by the hopping of oxygen vacancies may occur [13], while clear anomaly was not observed in **Figure 3**.

2.2 $SrHfO_3$

Perovskite oxide $SrHfO_3$ is one of current topics as a 5d band insulator. $SrHfO_3$ has attracted attention as a high-k dielectric $SrHfO_3$-based device with ultralow leakage current density [17]. The structural and vibrational properties of $SrHfO_3$ were investigated using first-principles calculations [18]. In the cubic phase ($Pm\bar{3}m$), the phonon dispersion curves indicate the coexistence of the ferroelectric instability at the Brillouin zone-center Γ point and antiferrodistortive instabilities at the zone-boundary R and M points. The *Pnma* ground-state is stabilized by antiferrodistrotive instabilities, which enhances the successive condensation of three rotational modes and suppresses the ferroelectric distortion. Upon cooling from high-temperature, cubic $SrHfO_3$ undergoes the successive phase transition of a heterophonon sequence [19]. Powder neutron diffraction study reported that the structure of $SrHfO_3$ is

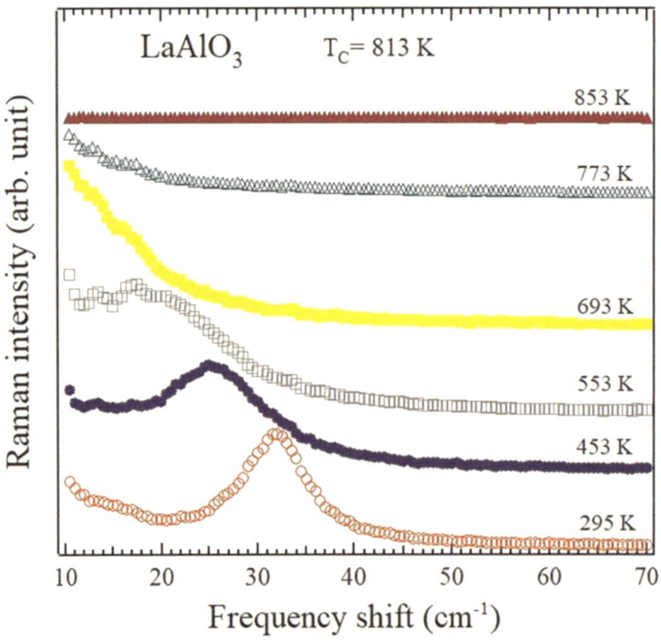

Figure 2.
The temperature dependences of Raman scattering spectra of E_g modes of a $LaAlO_3$ crystal in the low-frequency range observed at back scattering geometry.

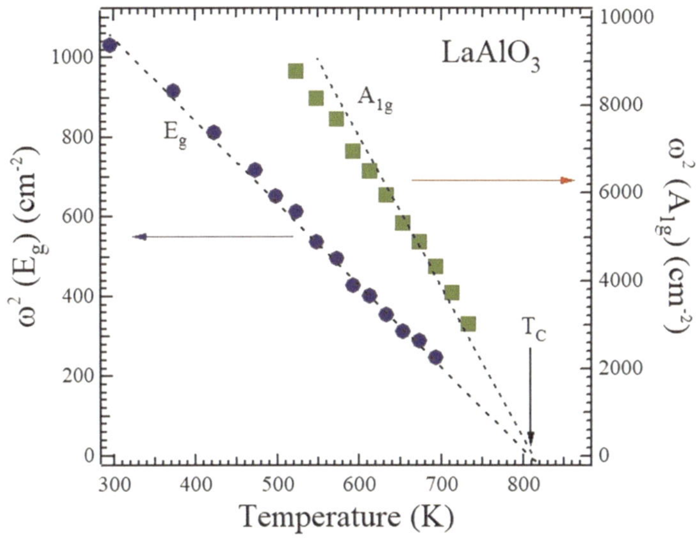

Figure 3.
The squared mode frequency of A_{1g} and E_g modes of a $LaAlO_3$ crystal as a function of temperature observed by Raman scattering.

orthorhombic (*Pnma*, Z = 8) from 300 K to approximately T_{oo} = 670 K. The high-temperature phase has a second orthorhombic structure (*Cmcm*, Z = 8). Upon further heating, the antiferrodistortive phase transition occurs and $SrHfO_3$ transforms into a nonpolar tetragonal structure (*I4/mcm*, Z = 4) at T_{ot} = 1023 K. At higher temperatures ~1360 K, another antiferrodistortive phase transition occurs into the ideal cubic perovskite structure ($Pm\bar{3}m$, Z = 1). The angle of rotation of the oxygen octahedron in the tetragonal phase is taken as the order parameter, and its temperature variation is consistent with a second-order phase transition [20].

In the orthorhombic (*Cmcm*) phase below 1023 K, the cell doubling occurs by the antiferrodistortive phase transition. A twofold degenerate zone-boundary soft mode in the high-temperature tetragonal phase (*I4/mcm*) changes into nondegenerate zone-center Raman active modes, which are the B_{1g} mode at 107 cm^{-1} and the A_g mode at 232 cm^{-1} and room temperature. The temperature dependence of Raman scattering spectra was reported in the large temperature range from 300 to 1023 K [21]. The Raman scattering spectrum, including two soft modes at room temperature, is shown in **Figure 4**. The temperature variation of the lowest frequency B_{1g} mode in the orthorhombic phase (*Pnma*) at 107 cm^{-1} and room temperature is shown in **Figure 5**. The mode shows softening of a nearly second-order phase transition toward T_{ot} = 1023 K. Upon heating, the softening of the A_g mode at 232 cm^{-1} and room temperature was also observed. The anomalous temperature dependence in mode frequencies and linewidth was observed in four intense Raman modes at 144, 166, 424, and 599 cm^{-1} in the vicinity of the orthorhombic to second orthorhombic phase transition temperature, T_{oo} = 670 K.

2.3 $(Na_{0.5}Bi_{0.5})TiO_3$

According to the environmental problem, lead-free piezoelectrics have attracted much attention. The ferroelectric $(Na_{0.5}Bi_{0.5})TiO_3$ (NBT) with perovskite structure is

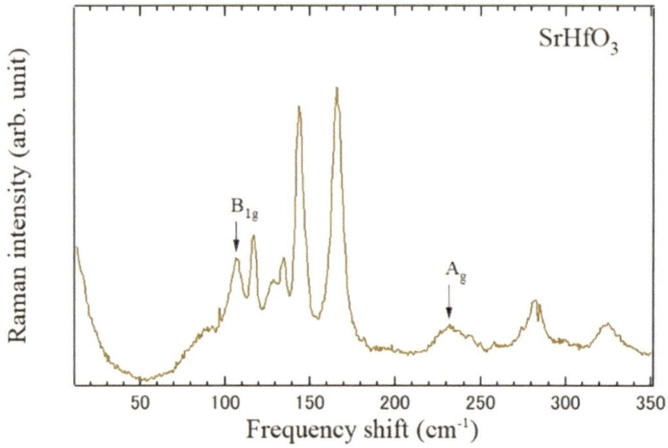

Figure 4.
Raman scattering spectrum of $SrHfO_3$. Arrows show the two soft modes of B_{1g} at 117 cm^{-1} and A_g at 232 cm^{-1}.

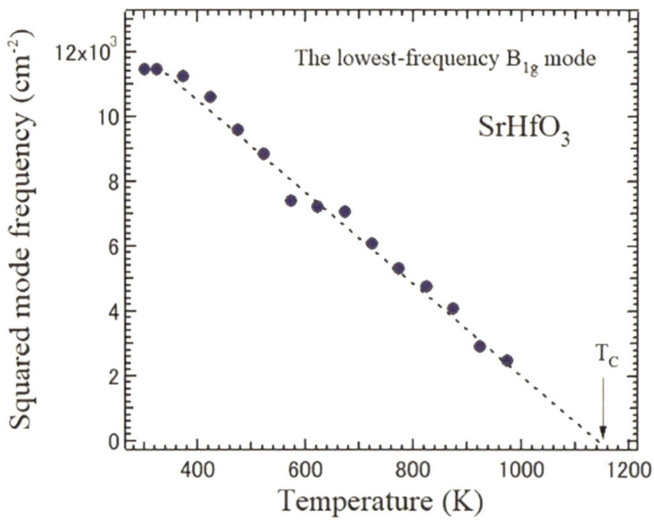

Figure 5.
The squared lowest-frequency B_{1g} mode of $SrHfO_3$ as a function of temperature observed by Raman scattering. The mode shows softening toward the nearly second-order orthorhombic to tetragonal phase transition at T_{ot} = 1023 K. Here, T_C is the extrapolated temperature.

one of the base materials in lead-free piezoelectrics. NBT (band gap, 3.1 eV) with a valence band formed by Bi(III) was also found as a new photocatalyst for solar water splitting [22]. In contrast to Pb-based perovskite relaxor ferroelectrics, pure NBT undergoes a successive phase transition with high diffusivity, and the remarkable frequency dispersion of a dielectric constant is observed only below a dielectric maximum temperature, T_m. These physical properties and the mechanism of the large piezoelectric effect near the MPB composition have not been clarified yet [23]. The structure of NBT is rhombohedral with space group *R3c* at room temperature. In the pseudocubic cell, the ion displacements occurs along the [111] direction and the out-of-phase $a^-a^-a^-$ octahedral tilting also occurs. Monoclinic or rhombohedral/

monoclinic mixtures were also reported at room temperature. In a ferroelectric phase of NBT at room temperature, the remnant polarization is P_r = 20–35 μC/cm^2 and the coercive field is $E_c \approx$ 70–90 kV/cm. On heating, it undergoes the following sequence of diffused structural transformations: The room temperature rhombohedral (*R3c*) phase transforms into a tetragonal (*P4bm*) phase at T_{RT} = 260–400°C, on further heating it transforms into a cubic ($Pm\bar{3}m$) phase at T_{TC} = 520–540°C [24].

Broadband inelastic light scattering spectra of NBT single crystals were measured in the frequency range between 2 and 1000 cm^{-1} to investigate the peculiar behavior of the polarization and the origin of diffusive dielectric anomaly [25]. The Brillouin scattering spectra between 0.2 and 20 cm^{-1} were measured by a tandem multi-pass Fabry-Perot interferometer [26]. The Raman scattering was measured by a triple-grating Raman spectrometer with the additive dispersion between 10 and 1000 cm^{-1}. In the light scattering spectrum between 2 and 200 cm^{-1}, quasielastic central peaks (CPs) were observed. Optical phonon modes were observed in the broadband inelastic light scattering spectrum between 50 and 1000 cm^{-1} (**Figure 6**).

Using the width of a broad CP, the inverse relaxation time was determined as a function of temperature. Upon cooling from the high temperature, it increases toward T_m. This result shows critical slowing down of the polarization fluctuations toward T_m as shown in **Figure 7(a)**. The another inverse relaxation time was also determined by the narrow CP. Its similar temperature dependence is similar to that of the broad CP, as shown in **Figure 7(b)**.

The LST relation holds for IR active LO phonon frequency ω_L, TO phonon frequency ω_T and dielectric constants, where ε_0 and ε_∞ are the dielectric constants ε for ω = 0 and ∞, respectively.

$$\frac{\varepsilon_0}{\varepsilon_\infty} = \prod_{j=1}^{n} \frac{\left|\omega_{Lj}\right|^2}{\left|\omega_{Tj}\right|^2} \tag{3}$$

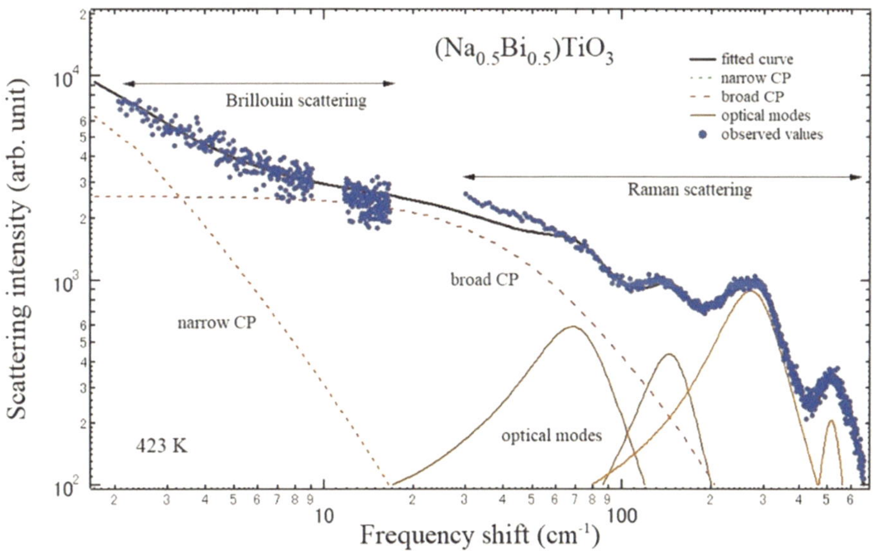

Figure 6.
Broadband inelastic light scattering was observed in a flux-grown NBT crystal at 423 K. It includes narrow and broad CPs and Raman active optical modes.

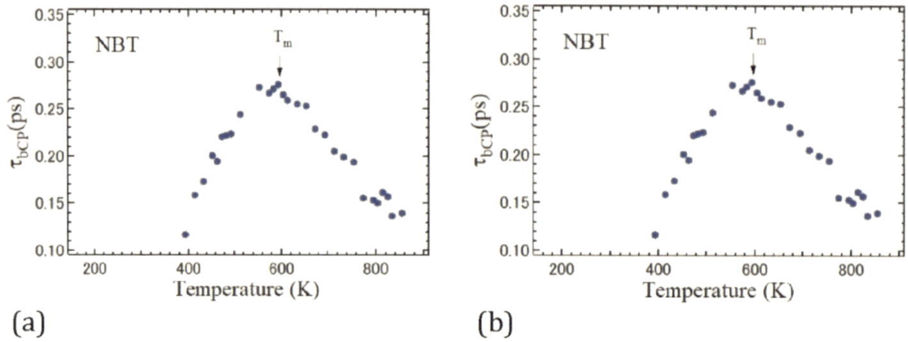

Figure 7.
Temperature variation of inverse relaxation times determined from the widths of (a) broad and (b) narrow central peaks. T_m *denotes the dielectric maximum temperature.*

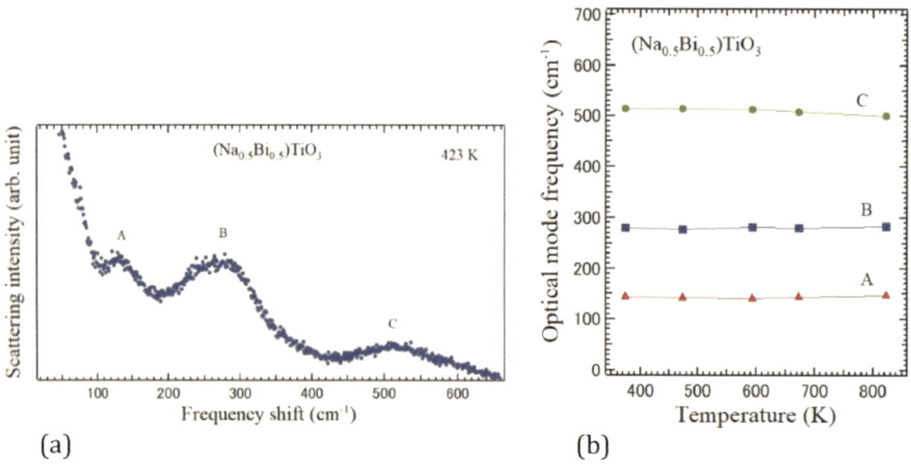

Figure 8.
(a) Raman scattering spectrum at 423 K and (b) optical mode frequencies of modes A, B, and C as a function of temperature.

The Raman scattering spectra at 423 K show three peaks of optical modes, as shown in **Figure 8(a)**. The three mode frequencies are nearly temperature independent, as plotted in **Figure 8(b)**. Therefore, the dielectric maximum near T_m is not reproduced in e_0 by the LST relation of Eq. (3).

The generalized LST relation is used to describe the dielectric behavior. As the contribution of relaxation processes to a dielectric constant, relaxation times τ_{Th}, τ_{Lh} are included in the generalized LST relation [27].

$$\frac{\varepsilon_0}{\varepsilon_\infty} = \prod_{h=1}^{m} \frac{\tau_{Th}}{\tau_{Lh}} \prod_{j=1}^{n_\alpha} \frac{|\omega_{Lj}|^2}{|\omega_{Tj}|^2} \tag{4}$$

Three phonon mode frequencies ω_1, ω_2, ω_3 and two relaxation times τ_{nCP}, τ_{bCP} reproduce the static dielectric constant. The observed dielectric constant and the static

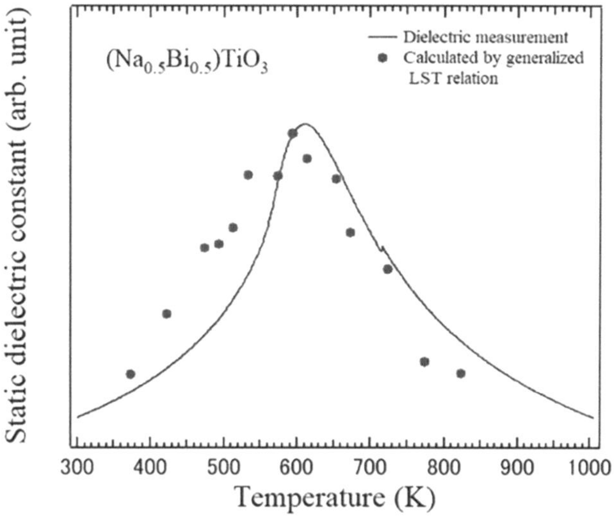

Figure 9.
Temperature dependences of observed dielectric constant and the static dielectric constant calculated by the extended LST relation.

dielectric constant calculated by the extended LST relation are shown as a function of temperature in **Figure 9**. The dielectric constant calculated by the generalized LST relation is in agreement with the observed dielectric constant. It means that the dominant role in a dielectric constant is the relaxation modes through the generalized LST relation, including the relaxation times determined by the narrow and broad CPs [25]. This result indicates that the diffusive dielectric anomaly near T_m was caused by the local polarization fluctuations. The temperature dependence of a dielectric constant obeys the extended Currie-Wiess (CW) law given by

$$\frac{1}{\varepsilon} - \frac{1}{\varepsilon_m} = \frac{(T - T_m)^\gamma}{C}, \tag{5}$$

where $\gamma = 1$ gives the normal CW law [23]. In NBT, the observed value is $\gamma = 1.44$, which indicates the highly diffusive nature. The origin of the diffusive nature of pure NBT is attributed to the random fields related to the random occupancy of Bi and Na ions at the A site.

3. Bismuth layer-structure ferroelectrics

The bismuth layer-structured ferroelectrics (BLSFs) have been extensively studied since the 1960s. The chemical formula is $Bi_2A_{m-1}B_mO_{3m+3} = (Bi_2O_2)^{2+}(A_{m-1}B_mO_{3m+1})^{2-}$, where n is natural number, A = Na^{1+}, K^{1+}, Ba^{2+}, Bi^{3+} and B=Fe^{3+}, Ti^{4+}, Nb^{5+}, W^{6+}. The $(Bi_2O_2)^{2+}$ denotes a bismuth oxide layer, and the $(A_{m-1}B_mO_{3m+1})^{2-}$ denotes a pseudo-perovskite layer, where m and m − 1 are the numbers of oxygen tetrahedra and of pseudo-perovskite units in a pseudo-perovskite layer, respectively [28, 29]. In BLSFs, a pseudo-perovskite layer is sandwiched between bismuth oxide layers. In the most ferroelectric phase transitions of BLSFs, underdamped soft modes were observed except the Ba compound, in which Ba with a large ionic size occupies the A-site in a pseudo-perovskite layer. Such a difference is known also in ferroelectric phase transitions of $ATiO_3$ (A = Ca, Sr., Pb, and Ba) with the perovskite structure.

3.1 Bismuth titanate

Among the BLSFs, bismuth titanate, $Bi_4Ti_3O_{12}$ (BIT) is one of the important key materials for electro-optic devices and non-volatile ferroelectric random access memories with low fatigue [30]. The charge redistribution of a spatially differentiated ferroelectric BIT single crystal was reported for photocatalytic overall water splitting [31]. Recently, the high-temperature phase transformation of the ferroelectric layered perovskite BIT has been re-examined by high-resolution powder neutron diffraction [32]. It was reported that BIT undergoes a successive phase transition upon cooling from the first tetragonal phase (space group *I4/mmm*) to the intermediate second tetragonal phase (space group *P4/mbm*) at 850–1000°C. For further cooling, the first-order ferroelectric phase transition occurs and transforms into the polar monoclinic phase (space group *B1a1*) at T_C = 675°C. The spontaneous polarization appears nearly along the *a*-axis, but it is inclined about 4.5° toward the *c*-axis in the *b*-plane. The *a*-axis component of a spontaneous polarization is $p_1 = 50$, and that of *c*-axis component is $p_3 = 4\ \mu C/cm^2$ at room temperature. Free energy of BIT is given by

$$F(p_1, p_3) = \frac{1}{2}a(T - T_0)p_1^2 + \frac{1}{4}\beta_1 p_1^4 + \frac{1}{2}\eta p_1^2 p_3^2 + \frac{1}{2}\alpha_3 p_3^2 + \frac{1}{4}\beta_3 p_3^4 + \frac{1}{6}\gamma\left(p_1^2 + p_3^2\right)^3. \quad (6)$$

Here, the phase transition induced by the bi-quadratic coupling $\frac{1}{2}\eta p_1^2 p_3^2$ ($\eta < 0$) is called the triggered transition [33]. BIT undergoes a phase transition upon cooling from a paraelectric phase with $p_1 = 0$, $p_2 = 0$, and $p_3 = 0$ into a ferroelectric phase with $p_1 \neq 0$ $p_2 = 0$, and $p_3 \neq 0$ at $T_C < T_0$.

Optical phonon modes with A'(*x,z*) and A"(*y*) symmetries are Raman and infrared active, where the *b*(*y*)-axis is perpendicular to a mirror plane. The symmetry of the soft mode is A'(*x,z*), which corresponds to p_1 and p_3. In the ferroelectric phase, the displacive nature was reported by the observation of the underdamped soft optic mode below T_C by Raman scattering [34]. Upon heating from room temperature, the A'(*x,z*) mode at 28 cm^{-1} remarkably softens toward T_C. The La doping effect on the soft mode was also investigated. In $La_xBi_{4-x}Ti_3O_{12}$, it was found that the lowest frequency soft mode shows a remarkable softening with an increase in the La content at room temperature [35].

Raman scattering spectra of A'(*x,z*) and A"(*y*) modes of a BIT crystal are shown in **Figure 10**. The lowest frequency A'(*x,z*) mode at 28 cm^{-1} is the ferroelectric soft mode [34]. The softening of the squared mode frequency toward T_C = 675°C is shown in **Figure 11**. The second lowest frequency A"(*y*) mode at 32 cm^{-1} is not a ferroelectric soft mode, and it does not show remarkable softening upon heating. Since A'(x,z) and A"(y) modes are also infrared active, the polariton dispersion relations of these modes were studied in a THz range [28].

The melt-quenched amorphous BIT was transformed into crystalline states by heat treatments. The two-step crystallization was studied by Raman scattering [36]. By annealing an as-quenched amorphous sample at 500°C, the intense lowest-frequency Raman peak was observed. This peak is similar to a boson peak of glassy materials.

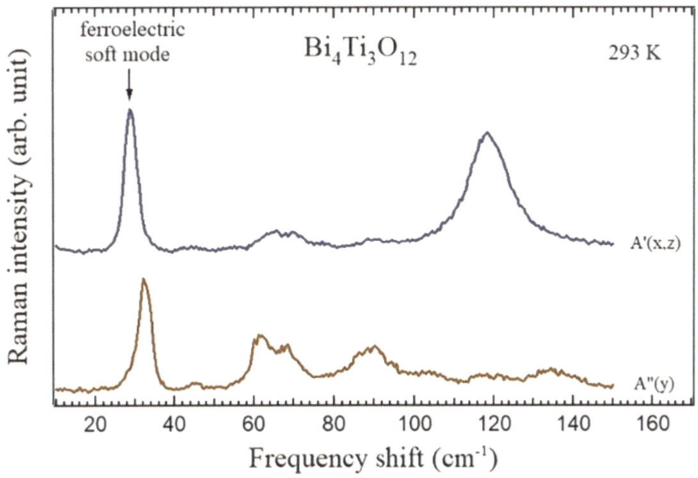

Figure 10.
Raman scattering spectra of a BIT crystal of A'(x,z) and A"(y) modes observed at the scattering geometries c(aa)$\bar{c}$ *and* c(ab)$\bar{c}$*, respectively, at 293 K. The lowest frequency A'(x,z) mode at 28 cm^{-1} is the ferroelectric soft mode.*

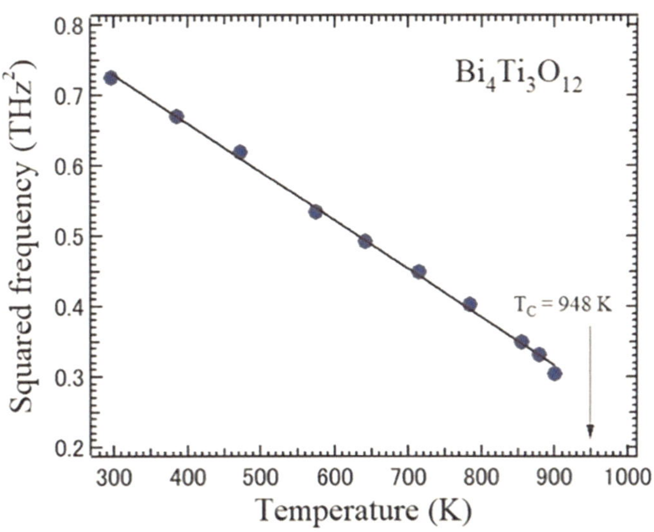

Figure 11.
Temperature variation of the squared lowest frequency soft mode with A'(x,z) symmetry of a BIT crystal.

A boson peak is related to an intermediate-range order. Using the peak frequency and transverse sound velocity, the medium-range correlation length of amorphous BIT is estimated to 7 nm. It is in agreement with the mean cluster size determined by the measurement of atomic force microscopy. The two-step crystallizations at 608 and 830°C were observed on heating by differential thermal analysis. Such crystallization processes suggest the existence of a metastable state. The structures of crystallized samples after annealing at 770 and 1000°C were ascribed to the pyroclore and the layered perovskite structures, respectively. These results indicate that at first a three-dimensional (3D) pyroclore structure is crystallized from a 3D amorphous state, and secondly, it changes into a pseudo-2D layered perovskite structure.

3.2 Strontium bismuth tantalate and its family

Recently, in bismuth layer-structured ferroelectrics $Bi_2A_{m-1}B_mO_{3m+3}$, the compounds of m = 2 have become to be the most important for the application to ferroelectric memory [37]. $SrBi_2Ta_2O_9$ and $SrBi_2Nb_2O_9$ are isostructural and crystallize in the polar $A2_1am$ space group at room temperature [38]. At high temperature, they crystallize in the prototypic *I4/mmm* structure as same as BIT. The lattice instability related to ferroelectric phase transitions of this family has been studied by Raman scattering in the temperature range from 90 to 1200 K. The softening of the lowest frequency soft mode of $SrBi_2Ta_2O_9$, at 27 cm^{-1} and room temperature, and the increase of the damping factor were observed toward T_C = 608 K [39]. The composition dependence of the low-frequency modes has also been investigated in its family $ABi_2M_2O_9$ (A = Ba, Ca, Sr., M = Nb, Ta) and Bi_3MTiO_9 (M = Nb, Ta) [40, 41]. The substitution of the A site causes remarkable change of low-frequency Raman spectra, while the effect of the B site is small in the spectra. The soft modes of these compounds mostly show softening toward T_C, except $BaBi_2M_2O_9$ (M = Nb, Ta), in which a large Ba ion occupies the A site of a pseudo-perovskite layers. The result has similarity with the perovskite ferroelectric $ATiO_3$ (A = Ca, Sr., Ba, Pb), in which only the soft mode of $BaTiO_3$ is overdamped [3].

When Ca ions substitute Sr. ions of SBT with a small ionic size, the soft mode frequency increases. The temperature dependences of the low-frequency Raman scattering spectra of $CaBi_2Ta_2O_9$ is shown in **Figure 12(a)** [42]. Upon heating from room temperature, the lowest frequency mode at 31 cm^{-1} remarkably softens. It is found that squared soft mode frequency is proportional to temperature, as shown in **Figure 12(b)**. The extrapolated temperature of zero frequency is higher than T_C = 1206 K. This fact indicates that the ferroelectric phase transition has a displacive nature and is of the first order.

In contrast with the substitution by Ca ions, for the substitution by Ba ions with a large ionic size, the soft mode peak was not observed at room temperature. These facts indicate that as the A-O bond distance decreases, T_C increases, and the damping factor of a soft mode decreases. For the A site occupancy by alkali earth metals (Ca, Sr., Ba), the bond to surrounding oxygen is purely ionic, and the ferroelectric Curie temperature increases as the ionic radius of the A site ion decreases. Consequently, the displacement of a B-site ion is controlled by the change of ionic radius at the A site. While for the A site occupancy by Pb ions, the soft mode spectrum is highly distorted, because the covalency of the Pb-O bond has a large dynamical effective charge at the A site, and the large off-center of a Pb ion also contributes to the spontaneous polarization. Therefore, such a large change in the soft mode pattern induces the line shape on soft mode spectra.

4. Ferroelectrics with perovskite slab structure

Ferroelectric compounds with a perovskite slab structure (or perovskite-derived layered structure), $A_2B_2O_7$ (A = Ca, La, Nd, Sr., B=Nb, Ta, Ti) family, such as $Sr_2Nb_2O_7$, $Ca_2Nb_2O_7$, $La_2Ti_2O_7$, have attracted great attention as the highest T_C ferroelectric compound for the application to electrooptic devices [43]. Recently, $Sr_2Nb_2O_7$ and $Sr_2Ta_2O_7$ have shown better photocatalytic performance compared to non-layered materials of similar composition [44]. Structural phase transitions of $A_2B_2O_7$ compounds with perovskite slab structure show a displacive nature, and underdamped soft modes were observed. This fact may be attributed to the two-dimensional free space between pseudo-perovskite layers.

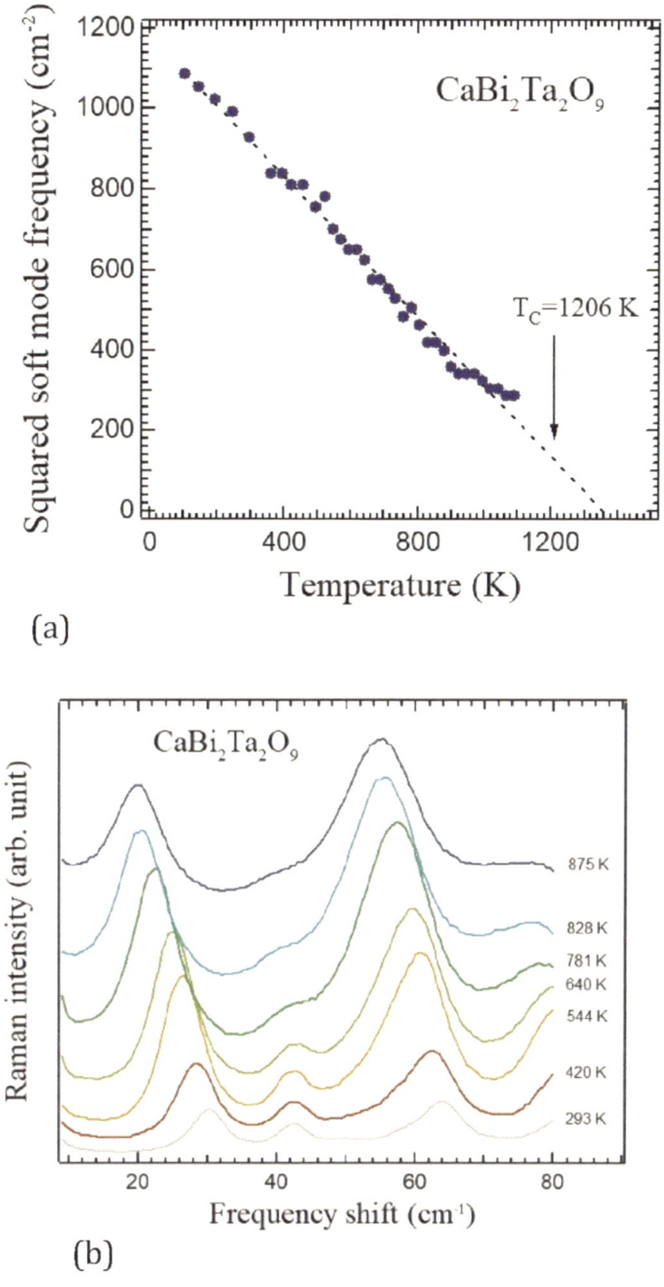

Figure 12.
The temperature variation of (a) the low-frequency Raman scattering spectra and (b) he squared soft mode frequency of $CaBi_2Ta_2O_9$.

$Sr_2Nb_2O_7$ undergoes a ferroelectric phase transition at T_C = 1342°C from an orthorhombic paraelectric phase (*Cmmm*) to a ferroelectric phase (*Cmc2₁*) with a spontaneous polarization along the c-axis [45]. In a Raman scattering spectrum at room temperature, the second lowest frequency optical mode with $A_1(z)$ symmetry at 56 cm^{-1} shows a remarkable softening toward T_C [46]. Since the soft mode is

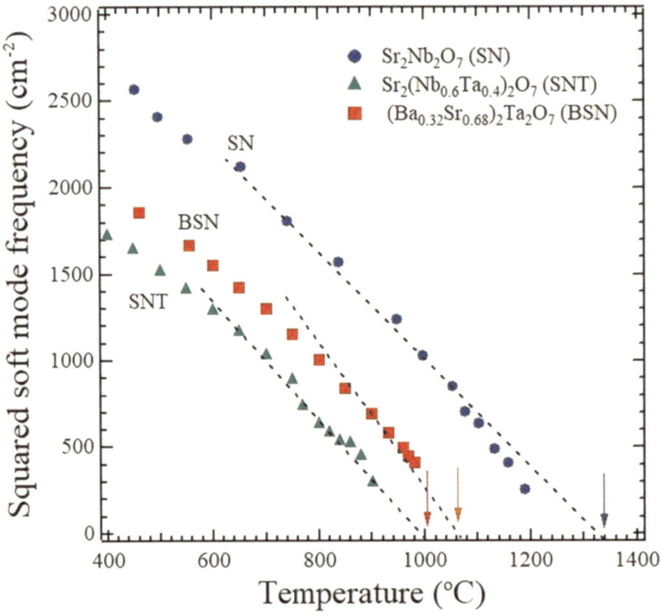

Figure 13.
Temperature dependences of squared mode frequencies of ferroelectric soft modes with $A_1(z)$ symmetry on $Sr_2(Nb_{1-x}Ta_x)_2O_7$ with x = 0.00, 0.20, 0.40, 1.00 and $(Ba_ySr_{1-y})_2Nb_2O_7$ (y = 0.00, 0.32) crystals.

underdamped over the large temperature range, the ferroelectric phase transition of $Sr_2Nb_2O_7$ can be the typical displacive type. Such underdamped soft modes were also observed in $Sr_2(Nb_{1-x}Ta_x)_2O_7$ and $(Ba_ySr_{1-y})_2Nb_2O_7$ crystal [47, 48]. **Figure 13** shows the variation of softening in the $Sr_2Nb_2O_7$ family. These substitution in the A- and B-sites causes a big variation of T_C as shown in **Figure 14** [47, 48].

The tunability of T_C is very important not only in fundamental physics but also in the industrial applications of ferroelectric materials. It can control various functionalities such as dielectric, pyroelectric, piezoelectric, electrooptic, and thermal properties. In various piezoelectric and ferroelectric materials, the effect of various substitutions on the A- and B-sites in ferroelectric compounds has been extensively studied. However, the colossal tunability of T_C over a large temperature range was not yet reported except $Sr_2Nb_2O_7$ family (**Figure 14**). The excellent tunability of T_C observed in $Sr_2(Nb_{1-x}Ta_x)_2O_7$ is greater than 1400 °C.

5. Dispersion of phonon-polariton in ferroelectrics

The coupled excitation between photons and other elementary excitations is called a polariton [49]. The frequency of polaritons shows the remarkable vs. wavevector dependence. Such a polariton dispersion relation is very important in fundamental sciences to understand the elementary excitations in condensed matter, such as phonons, plasmons, and excitons. The coupled excitation between infrared active phonons and photons is known as phonon-polariton, which has rich of application on various light sources such as a tunable laser and tunable terahertz radiation source [4]. The dispersion of polariton is remarkable only in the very small wavevectors in comparison with those of visible light. Consequently, the polariton dispersion is

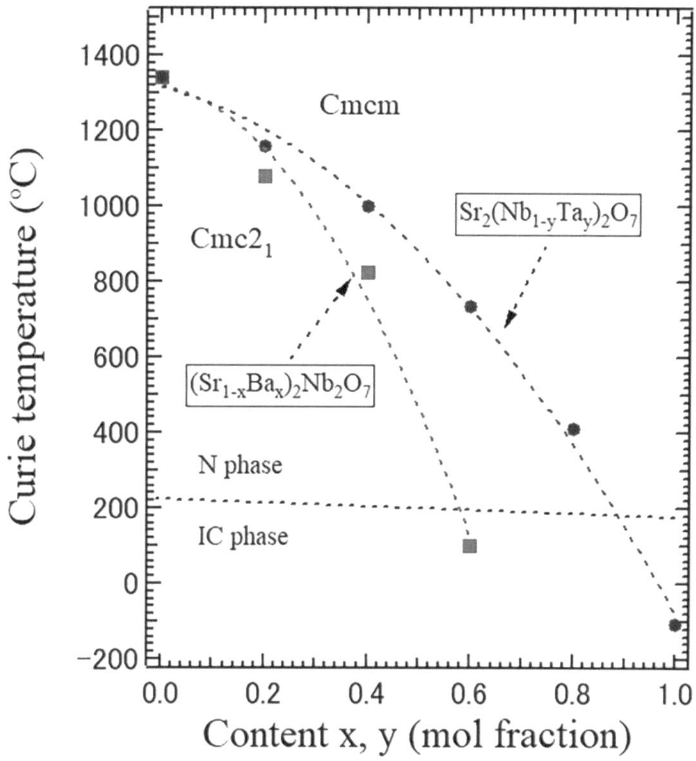

Figure 14.
Tunability of ferroelectric Curie temperatures of $(Ba_xSr_{1-x})_2Nb_2O_7$ and $Sr_2(Nb_{1-x}Ta_x)_2O_7$. The dotted line shows the incommensurate phase transitions of $Sr_2(Nb_{1-x}Ta_x)_2O_7$.

negligible in the standard Raman scattering at back or right-angle scattering geometry cannot detect the wavevector dependence of polariton frequency [28]. Till now, in most of the experimental studies on polaritons, the dispersion relation of phonon-polariton has been studied using forward Raman scattering [50] and impulsive stimulated Raman scattering (ISRS) [51].

In light scattering, the magnitude of the polariton wavevector q is determined by the Eq. (7) using the magnitude of wavevectors of the incident light k_i, scattered light k_s, and the scattering angle θ (**Figures 1** and **15**) [52].

$$q^2 = {k_i}^2 + {k_s}^2 - 2k_i k_s \cos\theta. \qquad (7)$$

For the determination of dispersion relation of polaritons, the scattering angle dependence of Raman spectra is required [53].

Lithium tantalate, $LiTaO_3$, with the trigonal space group *R3c* is technologically important by pyroelectric, piezoelectric, and electrooptic effects. In the trigonal symmetry, $A_1(z)$ and $E(x,y)$ symmetry modes are infrared and Raman active. **Figure 16** shows the forward Raman scattering spectrum of $A_1(z)$ symmetry observed at the scattering angle θ=0.54 ~ 0.64° in the scattering geometry $a(cc)a + \Delta b$ [53]. The lowest frequency $A_1(z)$ polariton was observed at 69 cm^{-1}, which is coupled to the lowest frequency TO mode at 209 cm^{-1}.

Figure 17 shows the Raman scattering spectrum of $E(x,y)$ modes at the scattering angle θ = 0.58 ~ 0.83° in the forward scattering geometry $a(cc)a + \Delta b$. The lowest

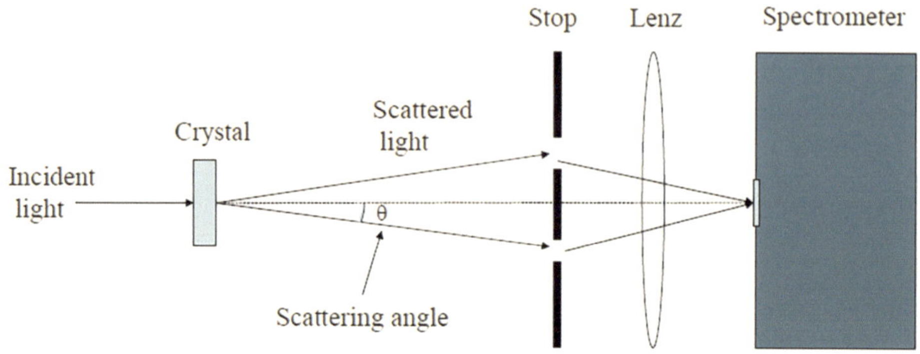

Figure 15.
Forward Raman scattering geometry to observe phonon-polaritons.

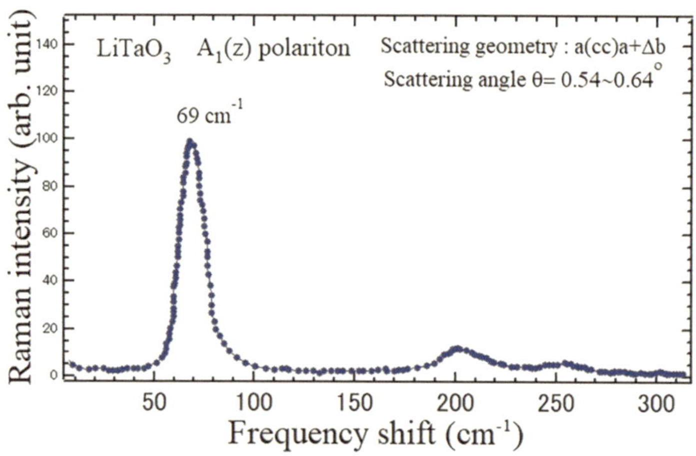

Figure 16.
Forward Raman scattering spectrum of A_1(z) modes observed at the scattering angle θ = 0.54 ~ 0.64°. The A_1(z) polariton was observed at 69 cm^{-1}.

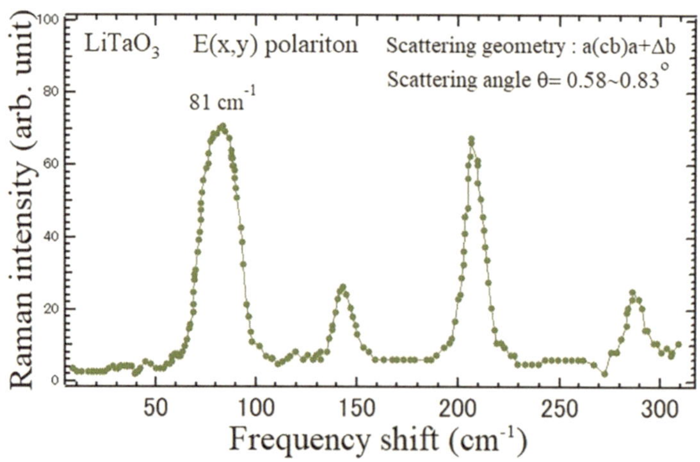

Figure 17.
Forward Raman scattering spectrum of E(x,y) modes at the scattering angle θ = 0.58 ~ 0.83° in the scattering geometry a(cb)a + Δb. *The lowest frequency E(x,y) polariton was observed at 81 cm^{-1}.*

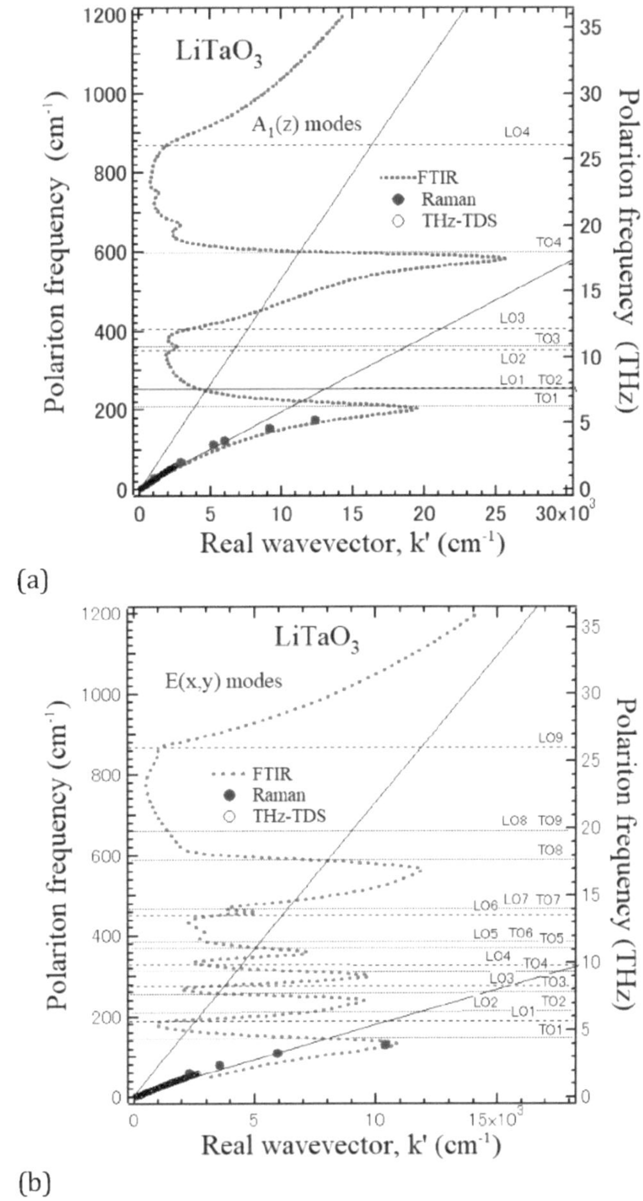

Figure 18.
Polariton dispersion of a LiTaO$_3$ crystal was determined in the broad frequency range from 10 to 1200 cm^{-1}. Polariton dispersion relations (a) A$_1$(z) and (b) E(x,y) modes determined by forward Raman scattering, THz-TDS, and FTIR measurements.

frequency E(x,y) polariton was observed at 81 cm^{-1}, which is coupled to the lowest frequency TO mode at 143 cm^{-1}.

By the combination with the observed polariton dispersion relations determined by forward Raman scattering, THz-TDS, and FTIR spectroscopy [4, 54], the broad-band polariton dispersion relation with A$_1$(z) and E(x,y) symmetry between 10 and 1200 cm^{-1} was determined as shown in **Figure 18(a)** and **(b)**, respectively.

The observed values of both symmetries by forward Raman scattering, THz-TDS, and FTIR measurements shows a good agreement within experimental accuracy.

6. Conclusions

Ferroelectric materials have garnered considerable interest for advanced electronic and optoelectronic applications. In metals, semi-metals, semiconductors, and superconductors, carrier electrons play a dominant role. In contrast, ferroelectric materials are insulators with a wide energy gap, and the lattice dynamics play a dominant role in ferroelectricity. Soft optic modes on the displacive type and fast relaxation processes (central peaks) on the order-disorder type have been investigated by terahertz Raman spectroscopy. This chapter gives new insights into the terahertz Raman spectroscopy of ferroelectrics. The temperature dependences of ferroelectric soft modes and central peaks in the large temperature range, and the broadband phonon-polariton dispersion are reviewed in the technologically important ferroelectric materials.

Author details

Seiji Kojima
Department of Materials Science, University of Tsukuba, Tsukuba, Ibaraki, Japan

*Address all correspondence to: kojima@ims.tsukuba.ac.jp

References

[1] Curie J, Curie P. Développement par compression de l'électricité polaire dans les cristaux hémièdres à faces inclinées. Bulletin de minéralogie. 1880;**3–4**:90-93

[2] Valasek J. Piezoelectric and allied phenomena in Rochelle salt. Physics Review. 1920;**15**:537-538. DOI: 10.1103/PhysRev.17.475

[3] Kojima S. Vibrational spectroscopy of perovskite ferroelectrics. Solids. 2024;**5**(4):593-616. DOI: 10.3390/solids5040040

[4] Kojima S. Broadband terahertz spectroscopy of phonon-polariton dispersion in ferroelectrics. Photonics. 2018;**5**(4):55. DOI: 10.3390/photonics5040055

[5] Kojima S. Gigahertz acoustic spectroscopy by micro-Brillouin scattering. Japanese Journal of Applied Physics. 2010;**49**:07HA01. DOI: 10.1143/JJAP.49.07HA01

[6] Kojima S. 100th Anniversary of Brillouin Scattering. Basel: MDPI; 2025. DOI: 10.3390/books978-3-7258-4108-0

[7] Raman CV, Krishnan KS. A new type of secondary radiation. Nature. 1928;**121**:501. DOI: 10.1038/121501c0

[8] Fonteset A et al. Raman, hyper-Raman, hyper-Rayleigh, two-photon luminescence and morphology-dependent resonance modes in a single optical tweezers system. Physical Review E. 2005;**72**:012903. DOI: 10.1103/PhysRevE.72.012903

[9] Borchers TH et al. Terahertz-Raman spectroscopy for in situ benchtop monitoring of changes to extended, supramolecular structure in milling mechanochemistry. Chem. 2025;**11**(2):102319. DOI: 10.1016/j.chempr.2024.09.018

[10] Kojima S. Vibrational spectroscopy of lead-free potassium sodium niobate and related perovskite ferroelectrics. Electronics and Signal Processing. 2025;**1**:0002. DOI: 10.55092/esp20250002

[11] Sun X. Anisotropic photoresponse behavior of a $LaAlO_3$ single-crystal-based vacuum-ultraviolet photodetector. Nanoscale. 2022;**14**:16829-16836. DOI: 10.1039/D2NR04552D

[12] Muhammad RM et al. A review on perovskite lanthanum aluminate ($LaAlO_3$), its properties and applications. Materials Research Express. 2019;**6**(11):112001. DOI: 10.1088/2053-1591/ab4629

[13] Hayward SA et al. Transformation processes in $LaAlO_3$: Neutron diffraction, dielectric, thermal, optical, and Raman studies. Physical Review B. 2005;**72**:054110. DOI: 10.1103/PhysRevB.72.054110

[14] Glazer AM. The classification of tilted octahedra in perovskites. Acta Crystallogr., Sect. B. 1972;**28**:3384. DOI: 10.1107/S0567740872007976

[15] Abrashev MV et al. Comparative study of optical phonons in the rhombohedrally distorted perovskites $LaAlO_3$ and $LaMnO_3$. Physical Review B. 1999;**59**:4146. DOI: 0.1103/PhysRevB.59.4146

[16] Fujita Y. Structural phase transition of ferroelastic $LaAlO_3$ studied by inelastic light scattering [Master thesis]. Tsukuba: University of Tsukuba; 2014

[17] Kim J et al. Low leakage in high-k perovskite gate oxide $SrHfO_3$. Advanced Electronic Materials. 2022;**9**(6):2201341. DOI: 10.1002/aelm.202201341

[18] Amisi S. First-principles study of structural and vibrational properties of $SrHfO_3$ compared to $SrZrO_3$. Computational Condensed Matter. 2019; **20**:e00383. DOI: 10.1016/j.cocom.2019.e00383

[19] Aizu K. General considerations of heterophone sequences of transitions. Journal of the Physical Society of Japan. 1977;**42**(2):424-432. DOI: 0.1143/JPSJ.42.424

[20] Kennedy BJ, Howard CJ, Chakoumakas BC. High-temperature phase transitions in $SrHfO_3$. Physical Review B. 1999;**60**:2972. DOI: 0.1103/PhysRevB.60.2972

[21] Singh MK et al. High-temperature phase transitions in $SrHfO_3$: A Raman scattering study. EPL. 2014;**107**:26004. DOI: 10.1209/0295-5075/107/26004

[22] Watanabe K et al. Highly crystalline $Na_{0.5}Bi_{0.5}TiO_3$ of a photocatalyst valence-band-controlled with Bi(iii) for solar water splitting. Chemical Communications. 2021;**57**:323-326. DOI: 10.1039/D0CC07371G

[23] Kojima S. Micro-brillouin scattering of normal and relaxor ferroelectrics. Ferroelectrics. 2011;**414**(1):12-23. DOI: 10.1080/00150193.2011.577282

[24] Jagło G et al. New insights into structural, optical, electrical and thermoelectric behavior of $Na_{0.5}Bi_{0.5}TiO_3$ single crystals. Scientific Reports. 2025;**15**:2733. DOI: 10.1038/s41598-025-86625-4

[25] Onda Y, Tsukada S, Kojima S. Raman and Brillouin scattering study of Lead-free ferroelectric (1-x)(Na0.5Bi0.5) TiO_3-xBaTiO_3 single crystals. Ferroelectrics. 2011;**412**(1):52-58. DOI: 10.1080/00150193.2011.542696

[26] Kojima S. 100th anniversary of Brillouin scattering: Impact on materials science. Materials. 2022;**15**:3518. DOI: 10.3390/ma15103518

[27] Chaves AS, Porto SPS. Generalized Lyddane-Sachs-teller relation. Solid State Communications. 1973;**13**(7):865-868. DOI: 10.1016/0038-1098(73)90386-4

[28] Kojima S. Terahertz Spectroscopy in Advanced Materials Science. London, UK: IntechOpen; 2023. DOI: 10.5772/intechopen.110850

[29] Subarao EC. A family of ferroelectric bismuth compounds. Journal of Physics and Chemistry of Solids. 1962;**23**(6): 665-676. DOI: 0.1016/0022-3697(62)90526-7

[30] Park BH, Kang BS, Bu SD, Noh TW, Lee J, Jo W. Lanthanum-substituted bismuth titanate for use in non-volatile memories. Nature. 1999;**401**:682-684. DOI: 10.1038/44352

[31] Jia G et al. Charge redistribution of a spatially differentiated ferroelectric $Bi_4Ti_3O_{12}$ single crystal for photocatalytic overall water splitting. Nature Communications. 2024;**15**:4746. DOI: 10.1038/s41467-024-49168-2

[32] Guo YY et al. Unexpected phase transition sequence in the ferroelectric $Bi_4Ti_3O_{12}$. IUCrJ. 2019;**6**:438-446. DOI: 10.1107/S2052252519003804IUCrJ

[33] Ishibashi Y. A phenomenological theory of the ferroelectric phase transition of $Bi_4Ti_3O_{12}$. Japanese Journal of Applied Physics. 1995;**34**(9S): 5475-5477. DOI: 10.1143/JJAP.34.5475

[34] Kojima S, Shimada S. Soft mode spectroscopy of bismuth Titanate single crystal. Physica B. 1996;**219&220**: 617-619. DOI: 10.1016/0921-4526(95)00830-6

[35] Anwar F, Ko JH, Kojima S, Lee SS, Jiang MS. Raman scattering study of A- and B-site substitutions in ferroelectric $Bi_4Ti_3O_{12}$. Journal of the Korean Physical Society. 2002;**41**:763-768. UCI : G704-000411.2002.41.5.008 (DOI is not provided.)

[36] Kojima S et al. Crystallization of amorphous bismuth titanate. Journal of Non-Crystalline Solids. 2001;**293–295**: 250-254. DOI: 10.1016/S0022-3093(01)00828-6

[37] Bozgeyik MS. Investigating structural features of Ba and Zr co-substituted strontium bismuth tantalate thin films. Bulletin of Materials Science. 2019;**42**:47. DOI: 10.1007/s12034-018-1723-y

[38] Pokhrel N, Nowadnick EA. Ferroelectric switching pathways and domain structure of $SrBi_2(Ta,Nb)_2O_9$ from first principles. Physical Review B. 2023;**107**:054108. DOI: 10.1103/PhysRevB.107.054108

[39] Kojima S. Optical mode softening of ferroelectric $SrBi_2Ta_2O_9$ and related bismuth layer-structured oxides. Journal of Physics: Condensed Matter. 1998;**10**: L327-L332. DOI: 10.1088/0953-8984/10/20/003

[40] Saitoh I, Kojima S. Soft phonon and bismuth content in ferroelectric $SrBi_2Ta_2O_9$. Physica B. 1999;**263-264**: 653-656. DOI: 10.1016/S0921-4526(98)01440-9

[41] Saitoh I, Kojima S. Softening of low-frequency modes in ferroelectric $Bi_3TiNb_2O_9$ and $Bi_3TiTa_2O_9$. Ferroelectrics. 1998;**217**(1):83. DOI: 10.1080/00150199808015024

[42] Kojima S. Raman spectroscopy of bismuth layer structured ferroelectrics. Ferroelectrics. 2000;**239**(1):55. DOI: 10.1080/00150190008213305

[43] Kojima S. Super-high Curie temperature ferroelectrics and superior T_C tunability: Lattice instability and ferroelectric phase transitions. Ferroelectrics. 2020;**569**(1):122-135. DOI: 10.1080/00150193.2020.1791657

[44] Bouri M et al. Oxygen evolution reaction activity of $Sr_2Ta_2O_7$ and $Sr_2Nb_2O_7$ surfaces. Journal of Physical Chemistry C. 2022;**126**(15):6556-6563. DOI: 10.1021/acs.jpcc.2c00649

[45] Nanamatsu S et al. Ferroelectric properties of $Sr_2Nb_2O_7$ single crystal. Journal of the Physical Society of Japan. 1971;**30**(1):300. DOI: 10.1143/JPSJ.30.300

[46] Kojima S et al. Polariton dispersion and infrared active soft mode in ferroelectric $Sr_2Nb_2O_7$. Journal of the Physical Society of Japan. 1981;**50**(9): 2787-2788. DOI: 10.1143/JPSJ.50.2787

[47] Kojima S, Tsukada S, Akishige Y. Inelastic light scattering of ferroelectric $Sr_2(Nb_{1-x}Ta_x)_2O_7$ crystals at high temperatures. Ferroelectrics. 2017; **512**(1):20-28. DOI: 10.1080/00150193.2017.1355148

[48] Hushur A, Akishige Y, Kojima S. Raman scattering study on ferroelectric $(Sr_{0.68}Ba_{0.32})_2Nb_2O_7$ ceramics. Materials Science and Engineering B. 2005;**120**: 45-49. DOI: 10.1016/j.mseb.2005.02.001

[49] Born M, Huang K. Dynamical Theory of Crystal Lattices. Oxford: Oxford University Press; 1954

[50] Henry CH, Hopfield JJ. Raman scattering by polaritons. Physical Review Letters. 1965;**15**:964-966. DOI: 10.1103/PhysRevLett.15.964

[51] Wiederrecht GP, Dougherty TP, Dhar L, Nelson KA, Leaird DE, Weiner AM. Explanation of anomalous polariton dynamics in $LiTaO_3$. Ferroelectrics. 1993;**150**:103-118. DOI: 10.1080/00150199308008698

[52] Kojima S, Nakamura T. Observations of low-frequency phonon-polariton in barium sodium niobate. Japanese Journal of Applied Physics. 1980;**19**:L609-L611. DOI: 10.1143/JJAP.19.L609

[53] Kojima S, Maczka M. Broadband phonon-polariton dispersion relation of ferroelectric $LiTaO_3$ crystals. Ferroelectrics. 2018;**533**(1):124-131. DOI: 10.1080/00150193.2018.1470825

[54] Kojima S, Mori T. Broadband terahertz time-domain spectroscopy of ferroelectric $LiTaO_3$: Phonon polariton dispersion. AIP Proceedings. 2014;**1627**: 52-57. DOI: 10.1063/1.4901657

Chapter 4

Unleashing the Power of Shape: How Metal Nanoparticle Morphology Supercharges Plasmonic and Polariton Effects to Revolutionize Raman Spectroscopy Sensitivity

Affi Nur Hidayah

Abstract

Raman spectroscopy identifies materials non-invasively at the molecular level by molecular vibrations induced by laser interaction with the material. Electrons in a molecular material are raised from their ground state into a virtual state by a laser's interaction, then return to the ground state and release a photon. Since the incident and emitted photons have different energies, the molecules vibrate, which Raman scattering detects. Nevertheless, the intrinsic weakness of Raman signals, particularly at low analyte concentrations, poses a major limitation. Therefore, metal nanoparticles with plasmonic properties are crucial for increasing the Raman signal mainly due to plasmonic effects, not because of chemical interactions that involve charge transfer. Plasmonic phenomena, such as surface plasmon polaritons (SPPs) and localized surface plasmon resonance (LSPR), are significantly influenced by essential factors including the metal nanoparticles' size, shape, and classification, together with the choice of substrate for surface-enhanced Raman spectroscopy (SERS). These phenomena collectively play pivotal roles in enhancing Raman signals. Moreover, the creation of "hot spots"—nanoscale gaps between adjacent nanoparticles—further intensifies local electromagnetic fields, significantly boosting sensitivity. Thus, systematic investigations into the interplay between nanoparticle morphology, plasmonic and polaritonic properties, and the resulting impact on Raman signal enhancement are essential for advancing the development of highly sensitive Raman-based detection platforms.

Keywords: metal nanoparticles, morphology, localized surface plasmon resonance, surface plasmon polariton, enhanced sensitivity of Raman spectroscopy

1. Introduction

In recent decades, Raman spectroscopy has received Raman spectroscopy has garnered significant interest in the research community as a vibrational spectroscopy technique more established than infrared spectroscopy due to its higher resolution arising from wavelength excitation in the visible and near-infrared ranges, enabling more precise identification of molecular changes [1, 2]. Vibrational spectroscopy is a technique that investigates molecular vibrations through the interaction of light, specifically electromagnetic radiation of a certain frequency, with materials. The vibrational properties of chemical molecular bonds define the nature of the chemical itself, while light-induced transitions between various vibrational frequencies can be spectroscopically examined to reveal details about particular molecules in a sample [3]. Raman spectroscopy, a non-destructive technique, can identify and examine biological, chemical, and physical samples [4] for diverse applications, including food detection [5], analysis of proteins such as nails, hair, skin, bones, gallstones, and kidney stones [6], forensics [7], environmental science [8], archaeology [9], biomedicine [10], and defense [11].

Cross-section Raman is around 10^{-29} cm^2/sr, resulting in a feeble Raman signal [12]. Raman scattering exhibits a cross-section value around 10^6 to 10^{14} times inferior to infrared absorption and fluorescence mechanisms; hence, it diminishes the sensitivity of Raman detection in low concentrations. A significant obstacle impedes the advancement and broader implementation of Raman [13]. It is essential to enhance Raman sensitivity. Weak Raman signals can typically be boosted by multiple magnitudes of orders by surface-enhanced Raman spectroscopy (SERS). The phenomenon initially documented in 1974 by Freischmann et al. by the examination of pyridine molecules adhered to the silver electrode's surface [14]. Van Duyne et al. identified that the Raman signal on a textured silver electrode is 10^5 to 10^6 times more intense than the Raman signal in the solution [15], as proposed by Moskovits in 1979 that the Raman signal would amplify in molecules adsorbed on noble metals [16]. The SERS phenomenon, which amplifies the Raman signal, is anticipated to improve Raman spectroscopy's sensitivity and solve the issue of weak detection at low concentrations.

SERS boosts Raman signal by increasing local electric fields through conduction electron oscillations or surface plasmons (SPs) produced during photon interactions with metal-dielectric surfaces, such as noble metal thin films or metal nanoparticle surfaces [17]. Propagating surface plasmon polaritons (SPPs) on noble metals and non-propagating localized surface plasmon resonance (LSPR) on nanoparticles (NPs) are two types of SP [18]. SERS conditions greatly affect Raman signal enhancement, which is defined by the enhancement factor (EF), which averages 10^4 to 10^6, often 10^8, and sometimes 10^{14} [19]. Chemical and electromagnetic enhancements affect SERS EF enhancement. Electromagnetic effects significantly improve the main signal enhancement in SERS, and the most effective is up to 10^{10} which is influenced by the characteristics of the nanoparticles and the substrate surface. Meanwhile, CM contributes less with an increase reaching 10^2 to 10^4 depends on the charge transfer between nanoparticles and molecules, hence significantly influenced by the molecular type [20]. A comprehensive understanding of the signal enhancement process, particularly electromagnetic (EM) augmentation, is crucial and warrants additional investigation.

The SERS substrate is categorized into three types: colloidal, rigid, and flexible substrates. The system that uses nanoparticle solutions is predominantly affected by LSPR. In contrast, the SERS substrate, whether in a rigid or semi-rigid state, is influenced not only by LSPR—determined by the nanoparticle type—but also by the substrate type, where surface plasmon polaritons (SPPs) significantly amplify the Raman signal. The substrate, hotspot distance, size, shape, and circumstances around the nanoparticle affect electromagnetic (EM) properties [21, 22]. Choosing the right nanoparticle features, substrates, and wavelengths when making and measuring SERS substrates greatly affects how much the Raman signal can be strengthened.

2. Raman spectroscopy: Principles and limitations of Raman spectroscopy

The Raman effect was discovered a century ago, initiated by Smekal's 1923 prediction of inelastic scattering as a result of light and molecules interacting, using classical quantum theory [23]. Nonetheless, experiments conducted during that period did not provide a direct explanation for the Raman effect, as the quantum concept had not been fully comprehended. C.V. Raman and K.S. Khrishnan observed the scattering of secondary radiation with reduced frequencies in 60 liquids and vapors, subsequently designating this phenomenon inelastic scattering, or the Raman effect in 1928 [24]. Landsberg and Mandelstam unintentionally noticed a similar result in quartz and Iceland spar that year while investigating wavelength variations in scattering according to the Debye specific heat theory in solids [25]. The publication by Raman and Landsberg established the basis for the identification of the Raman effect subsequently facilitated advancements in quantum theory and significant techniques to more effectively, quickly, and non-destructively analyze chemical compounds both qualitatively and quantitatively. Molecular detection has primarily employed conventional techniques, including Gas Chromatography-Mass Spectrometry (GC-MS), High-Performance Liquid Chromatography-Mass Spectrometry (HPLC-MS), Liquid Chromatography-Mass Spectrometry (LC-MS), Vacuum-Assisted Liquid-Liquid Microextraction-Liquid Chromatography-Mass Spectrometry (VALLMEs-LC-MS), Solid Phase Extraction coupled with Liquid Chromatography-Mass Spectrometry (SPE-LC-MS), Orbitrap or Time-of-Flight (ToF) Mass Spectrometry, and Ultra-Performance Liquid Chromatography-Mass Spectrometry (UPLC-MS). These methods exhibit limitations due to the complex preparation process, which may extend over several days and necessitate extensive purification [26, 27].

The interaction of monochromatic electromagnetic radiation with a material result in transmittance, scattering, and absorption. Absorption requires photon energy matching between electronic energy levels. Conversely, scattering does not require enough energy. The virtual state area and electron cloud are altered when photon-molecule interaction. The polarization shift is also caused. Elastic scattering happens when photon energy equals entering photon energy, and the electrons return to the initial state. It is inelastic otherwise. Stokes scattering makes Raman scattering inelastic because the released photon has less energy than the incoming photon. The Jablonsky diagram in **Figure 1** shows that photon energy causes anti-Stokes scattering. "Raman shift," the energy differential between incoming and departing photons, is a diatomic molecule's vibrational state energy [29].

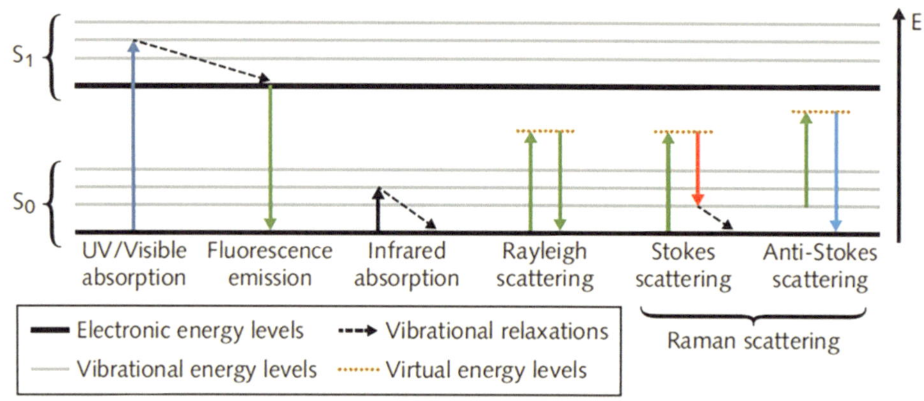

Figure 1.
Jablonski diagram illustrating Raman scattering vibrations [28].

Raman scattering is influenced by several things, which are represented as Raman intensity and characterized by the equation that follows:

$$I_R \propto I_0 v^4 N \left(\frac{\partial \alpha}{\partial Q} \right)^2 \tag{1}$$

A state's scattering molecules are denoted by N, polarizability by α, vibrational coordinate amplitude by Q, and input light intensity and frequency by Io. According to the right-hand final equation, Raman scattering requires photon contact polarizability modification [30].

Raman spectroscopy, a non-destructive technique, detects various compounds, however its scattering cross section is typically 10^{-16} cm^2 per molecule, making it weak [31]. Detecting low levels of contaminants, frequently in the ppm or ppb range, is difficult. Surface enhanced Raman spectroscopy or scattering uses metal nanoparticles' plasmonic effect to improve Raman sensitivity and overcome Raman scattering's limits [32].

3. Surface enhanced Raman scattering (SERS)

3.1 Basic theory

SERS amplifies the Raman signal by several orders from an analyte near metal nanoparticles as Au, Ag, Cu, or platinum. Surface (S) in SERS stands for Surface Spectroscopy, a technique that requires molecules to be on or near the surface. The term "Enhanced" (E) refers to the increase in signal generated by plasmon resonance on a metal substrate. "Raman" (R) denotes a measurement technique and the Raman signal of molecules. The final component, "S," can represent either spectroscopy or scattering, depending on whether the technology and its uses (spectroscopy) or the optical effect (scattering) are the main focus [33]. Light-illuminated metal nanoparticles can sustain surface plasmons, collective conduction electron oscillations at the metal-dielectric contact. The phenomenon known as localized surface plasmon resonance (LSPR) occurs when electrons locally vibrate coherently on or within nanoparticles, while propagating surface plasmons or polaritons (SPPs) propagate as longitudinal waves across the metal's surface, as shown in **Figure 2** [34].

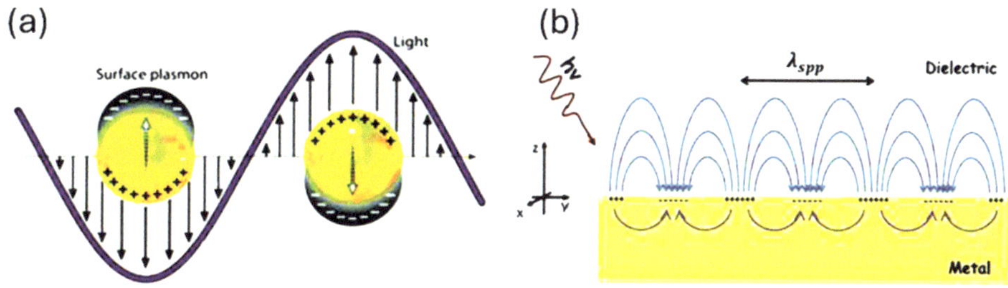

Figure 2.
(a) Localized surface plasmon resonance (LSPR) and (b) surface-plasmon polariton (SPP) images.

3.1.1 SERS enhancements

The increase in Raman signal from normal conditions is represented by the enhancement factor (EF), the most important factor in SERS. The correlation between SERS and Raman signal intensity is represented according to the subsequent equation, wherein the SERS signal intensity (I_{SERS}) is derived from multiplying the Raman intensity (I_{Raman}) by the enhancement factor (EF).

$$I_{SERS} = EF_{SERS} \text{ x } I_{Raman} = EF_{EM,SERS} \text{ x } EF_{CM,SERS} \text{ x } P_{Raman} \quad (2)$$

SERS enhancement is the augmentation of EM and CM contributions [13], shown in **Figure 3** [20].

3.1.1.1 Electromagnetic enhancement (EM)

Electromagnetic enhancement (EM) is affected by surface plasmon affected by characteristics of nanoparticles, such as their type (e.g., Au, Ag, or Cu), size, shape and the surrounding environment, specifically the substrate utilized. Electromagnetic (EM) effects constitute the primary factor enhancing. Placing the molecule in the SERS hot areas, between two or more adjacent nanoparticles (less than 10 nm), will increase the effectiveness of this signal augmentation. The interaction of a laser beam with nanoparticles or substrates coated with nanoparticles generates surface plasmons (either LSPR or SPP). Nanoparticle type, shape, size, and dielectric environment affect LSPR., while SPP is influenced not only by the nanoparticles' characteristics but also by the type of SERS substrate, whether colloidal substrate or a thin layer on a rigid substrate (silicon wafer, glass) or a flexible substrate (polymer, textile). The surface plasmon will produce an electromagnetic field that interacts with Raman scattering [35–37]. EM includes two unique contributions:

- Local field enhancement (or near field)

  Due to laser light amplification in hot spots, small spatial region, surface plasmon excitation strongly localizes. As a result, molecules in a hot spot experience an electromagnetic field that is far greater than what they would if the metal substrate were absent. Hot spots affect enhancement factor values, as seen in **Figure 4** [38].

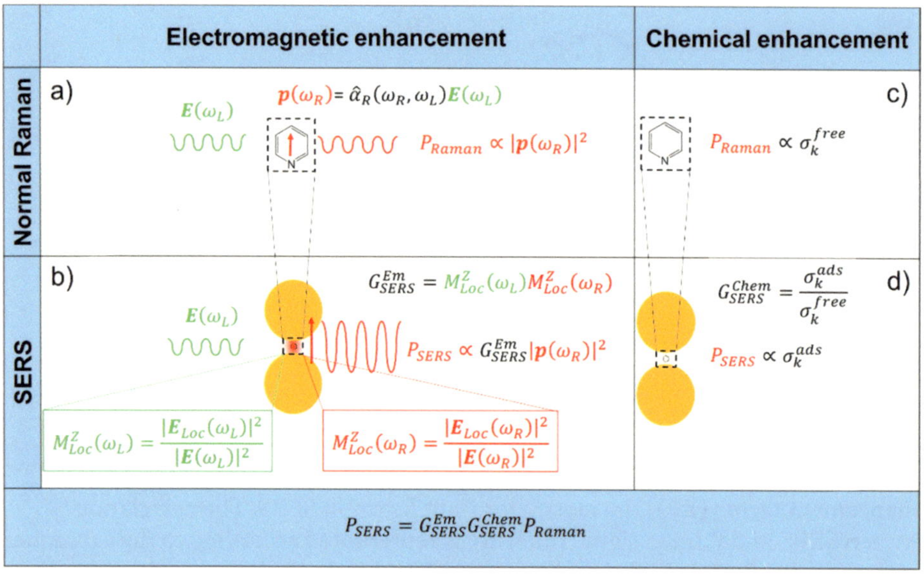

Figure 3.
The mechanism of Raman signal amplification using SERS; (a) in conventional Raman spectroscopy, $E(\omega_L)$, the electric field that oscillates at angular frequency ω_L incident onto the molecule, is defined by a tensor of Raman polarizability $\acute{\alpha}_R$ (ω_R, ω_L). Laser generates dipoles that oscillate at the Raman frequency, $p(\omega_R)$, and the dipole's magnitude equals the Raman power it radiates back, (b) denotes the electromagnetic enhancement in SERS. The molecule is near the plasmonic substrate, the incidence electric field E_{LOC} (ω_L) on the molecule typically exceeds the laser electric field E (ω_L), with the electric field enhancement quantified by $M_{LOC}^{Z}(\omega L)$. The plasmonic substrate amplifies the efficacy of dipole-induced Raman radiation, which is further amplified by $M_{LOC}^{Z}(\omega L)$. The aggregate of the electromagnetic wave enhancement factors, expressed as $|E|^4$, is denoted as $G_{SERS}^{Em} = M_{Loc}^{Z}(\omega_L)M_{Loc}^{Z}(\omega_R)$. (c) Denotes the chemical enhancement in regular Raman spectroscopy, represented by the free-space molecule vibrational modes defined by the cross-section (cross-section) *σ_k^{free}, while (d) signifies the SERS enhancement factor for chemical enhancement. Interaction of nanoparticles with molecules results in the adsorption of molecules onto nanoparticles, altering the molecular structure and therefore modifying the cross-section mode* (cross-section) (σ_k^{ads}). *The chemical enhancement is determined by the equation $G_{SERS}^{Chem} = \frac{\sigma_k^{ads}}{\sigma_k^{free}}$.*

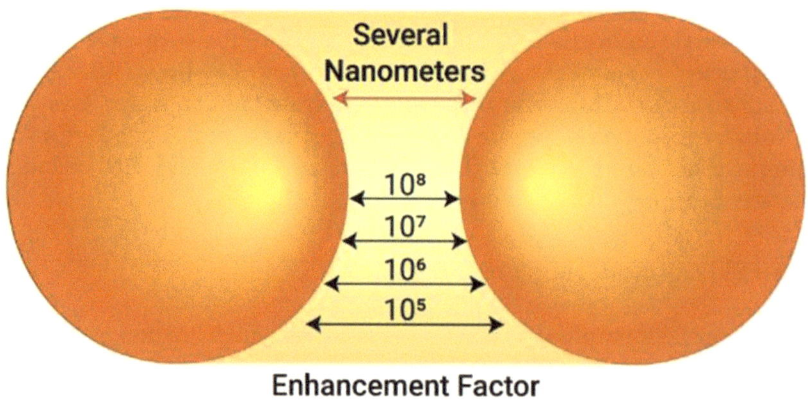

Figure 4.
Hot spot representation and enhancement factor.

- Enhanced re-radiation

  A metallic structure nearby affects a molecule's Raman power emission because a dipole's power depends on its surroundings [20].

3.1.1.2 Chemical enhancement (CM)

Chemical enhancement (CM) is the process by which analyte molecules and nanoparticles exchange charges, which increases due to the molecule's polarizability (the vibrational mode's Raman cross section) as a result of physicochemical interactions influenced by the molecule type. Chemical enhancement's role is often less significant than that of electromagnetic enhancement, ranging from 10^2 to 10^4. Chemical enhancement determines the spectrum pattern of SERS spectra, including the band intensity ratio and Raman shift [20]. Chemical enhancement is governed by two distinct mechanisms:

- The phenomenon of non-resonant chemistry

  When molecular orbitals are too far from the metal's Fermi level, the interaction between molecules and metals cannot form new electronic bands. However, the interaction can cause a multitude of changes in the molecules' geometry and electrical structure, as evidenced by changes in the Raman shift and vibrational mode intensity.

- Chemical processes involving resonant charge transfer

  Metal-molecule charge transfer (CT) states are produced when molecules and metals interact. It is possible to significantly amplify certain Raman modes, especially those connected to permitted electronic transitions (resonant Raman scattering), by utilizing laser source under resonance or pre-resonance conditions with these states [39, 40].

The representation of charge transfer in the interaction of metal nanoparticles and molecules and semiconductor nanoparticles is shown in **Figure 5** [38].

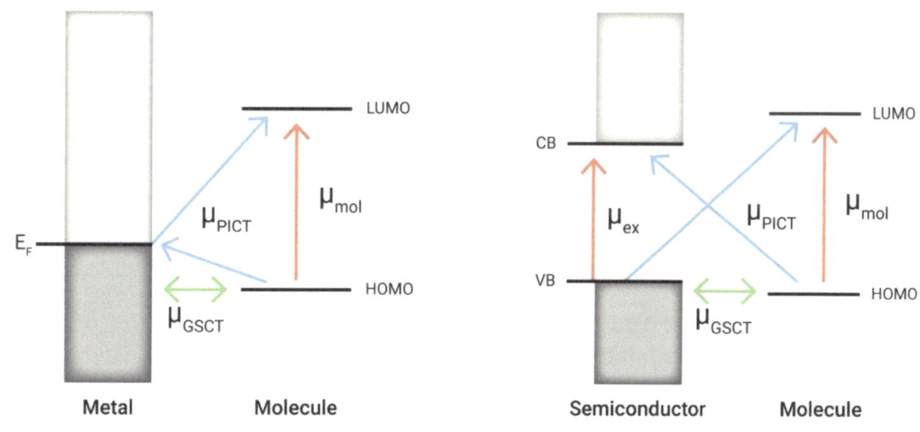

Figure 5.
Charge transfer between molecules and metal and semiconductor.

There are other factors than the augmentation factor that affect how sensitive the SERS technology is. Preparation and control of samples concerns significantly impact the efficiency of absorption; thus, analyte transfer from solution to substrate is a critical consideration. Several critical parameters influencing the improvement of Raman signal sensitivity must be considered while altering the production of SERS substrates and their measurement protocols:

- Comparison of SERS spectrum and Raman spectrum: At the same wavelengths, the SERS spectra and the regular Raman spectrum are usually comparable in most molecular analyses. Any discrepancy arises from several significant factors in the SERS process. The SERS spectrum is influenced by plasmon resonances, which affect how much the signal is boosted and the laser wavelength used for the measurement. Therefore, the difference between SERS and Raman spectra depends on the underlying resonance dispersion, which leads to enhancements arising from the symmetry of particular Raman modes as well as the orientation of particular molecules on the surface, resulting in various relative Raman peak intensities under SERS circumstances. Molecules may alter their "identity" during absorption and transform into surface complexes. This may lead to a minor displacement and widening of the Raman peak. It is also feasible to modify the intrinsic Raman intensity (cross-section) after absorption. Typically, these latter effects are categorized as chemical enhancements.

- Polarization effects: The Raman spectra and the SERS signal may diverge because polarization effects depending on the plasmonic resonance.

- The SERS continuum: SERS spectra are sometimes disturbed by background noise, which appears in most Raman spectra. The background is caused by impurities or residual intrinsic fluorescence.

- Photo-bleaching/photochemistry: Under typical non-SERS conditions, it is known that many SERS probes, including dyes, can experience photobleaching (at least when the excitation powers are large enough). Therefore, the fact that photobleaching happens under SERS circumstances is not surprising; decay of the SERS signal due to photobleaching is indeed observed experimentally and sometimes also appears in the SERS spectrum [10]. Moreover, the electromagnetic enhancement that produces the SERS signal has the potential to significantly impact the nature of the photobleaching, potentially leading to photochemical phenomenon, albeit the specifics are still unclear. Probe photostability (and related photochemical phenomena) may present issues depending on the specific probe, necessitating case-by-case analysis. However, it is essential to consider this when assessing SERS tests, particularly with dyes.

- Signal fluctuations: Additionally, SERS signals may show variations in intensity that differ from those of typical Raman. This can be attributed to changes in the configuration of the SERS substrate (e.g., colloidal Brownian motion), photobleaching, single-molecule sensitivity conditions, or combination [27, 33].

3.2 The effect of nanoparticles on Raman signal enhancement and its applications

3.2.1 Impact of nanoparticle classification

The interaction between adsorbed molecules and nanoparticles is crucial for enhancing Raman signals, and traditional SERS substrates often include either gold (Au), silver (Ag), or copper (Cu). Due to their air stability, Au and Ag are the most used SERS substrates, even though copper is more reactive. These three materials have visible and near-infrared wavelengths for their LSPR, where many Raman measurements are in this range. Other high-conductivity metals with plasmonic properties are Pt and Pd, which have LSPRs in the UV range [41]. Alkali metals like Pt, Rb, Li, Ga, Na, K, Cs, In, Al, and Rh are also being employed as SERS substrates. Since some of these substances react with air, stabilizing them might be challenging. Novel materials including graphene, semiconductors like ZnO and TiO_2, quantum dots, and other semiconductors are also used as SERS substrates [42].

The nanoparticle type influences the quality of Raman signal enhancement, as demonstrated by our prior research. This involved the detection and enhancement of the Raman signal from a cup of tea containing deltamethrin insecticide utilizing a liquid SERS system. Au, Au50Ag50, and Ag nanoparticles were directly combined with

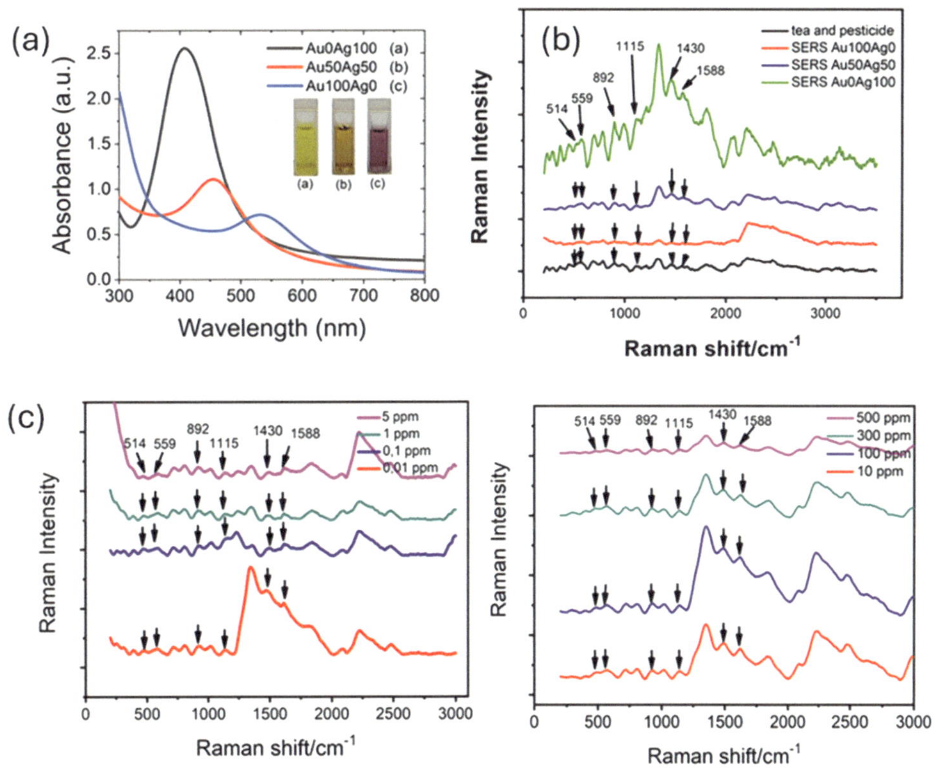

Figure 6.
(a) LSPR absorbance of Au-Ag nanoparticles, (b) Raman amplification of deltamethrin pesticide in Au-Ag liquid SERS, and (c) SERS of Ag solution deltamethrin pesticide identification LOD results.

the pesticide-containing tea drinks. The findings indicate that the Ag solution SERS achieved superior signal augmentation to the Au solution SERS and the Au50Ag50 alloy, as illustrated in **Figure 6**. The maximum Raman signal amplification in Ag nanoparticles is determined by the plasmonic characteristics of the most prominent Ag nanoparticles, whereas Au nanoparticles exhibit the least signal enhancement, and the Au50Ag50 alloy is positioned between Ag and Au. Nanoparticles employed in solution SERS are generated with a femtosecond laser that converts gold and silver salt solutions into nanoparticles [43].

Our additional research findings regarding the impact of nanoparticle types are presented, indicating that Ag-SERS substrates boost Raman signal of deltamethrin pesticide more than Au and Au-Ag substrates, as demonstrated in liquid SERS. SERS substrates are synthesized from colloidal Au, Au50Ag50 alloy, and Ag nanoparticles deposited on silicon wafer substrates. **Figure 7** illustrates the results of deltamethrin pesticides' increased Raman signal on the SERS substrate. Silver nanoparticles have the most pronounced plasmonic capabilities, while gold plasmonics are the least effective, and Au50Ag50 plasmonics fall between those of silver and gold, influencing the Raman signal's enhancement. The simulation indicates that silver nanoparticles have a greater electric field than gold and combined gold-silver nanoparticles. The simulation indicates that Ag plasmonics exhibit enhanced strength when subjected to a laser wavelength near its plasmonic resonance, with the LSPR of Ag at about 400 nm, Au at around 530 nm, and Au50Ag50 at about 450 nm. The electric field surrounding the nanoparticles is most optimal at a laser wavelength of 532 nm, as illustrated in **Figure 8** [44].

The effect of Raman signal enhancement on nanoparticle properties was also demonstrated by experiments conducted by Bozo et al., where R6G and MV's Raman spectra on Cu SERS substrates was able to be enhanced, but the results were lower when compared to the spectrum enhancement on CuPd and CuPt SERS substrates [45]. In the detection of bacteria with Cu substrates and Cu/Pd substrates, it was also shown that the type of nanoparticles would affect the increase in Raman signal, where the increase in Raman signal in bacterial detection on Cu/Pd substrates was higher than on Cu substrates [46].

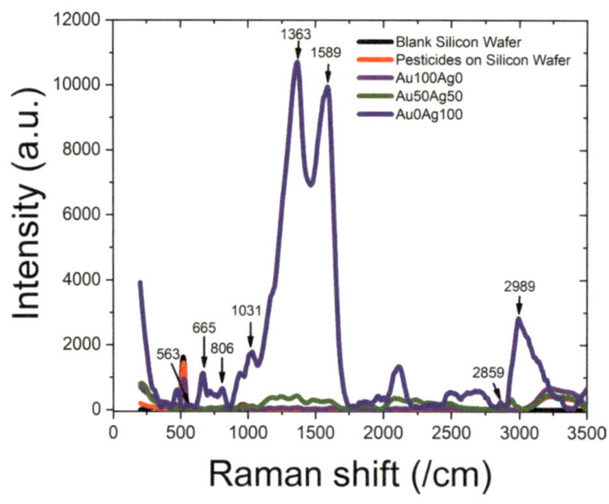

Figure 7.
Enhancement of Raman signals for deltamethrin insecticide on Ag SERS substrate.

DOI: http://dx.doi.org/10.5772/intechopen.1012206

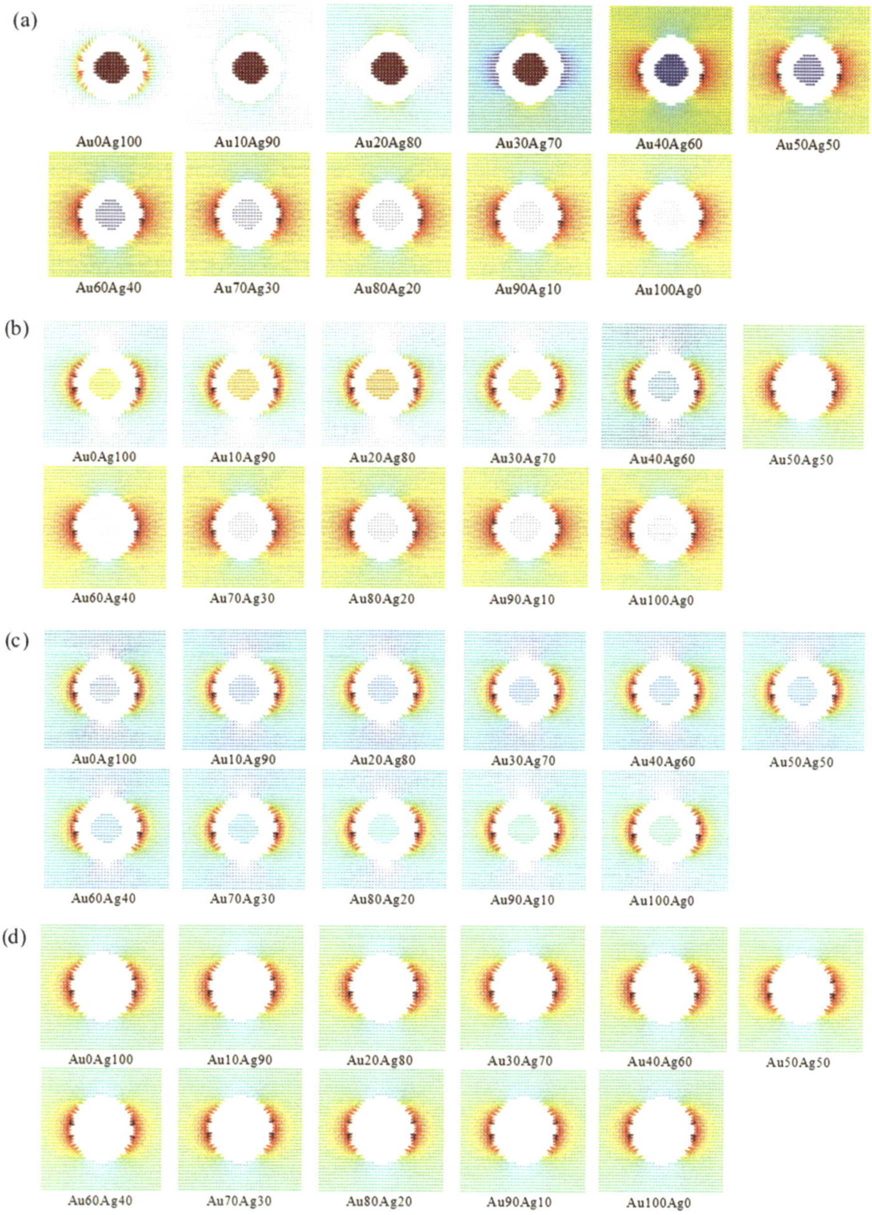

Figure 8.
Simulation of electromagnetic field dispersion in Au-Ag nanoalloys at an incident wavelength of (a) 400, (b) 450, (c) 532, and (d) 785 nm.

3.2.2 Impact of nanoparticle size

The plasmonic characteristics of LSPR are affected by the nanoparticles' size. When the nanoparticle gets bigger, the proximity between nanoparticles (hot spots) decreases, leading to an enhancement in electromagnetic (EM) effects and Raman signal. SERS intensity drops when particles scatter more light by inelastic scattering and absorb less light. As a result, figuring out the ideal nanoparticle size is crucial.

Kevin et al. documented this phenomenon in Ag nanoparticles at 20.5, 29.4, 33.0, 40.3, 59.3, and 69.5 nm in the R6G SERS signal, revealing the maximum SERS intensity was obtained at a size of 40.3 nm, identified as the saturation point, beyond which the SERS signal diminishes with increasing nanoparticle size. In their experiment, Soma et al. found that Ag nanoparticle size affected the greatest SERS signal enhancement. They utilized 532 nm laser ablation to synthesize nanoparticles of 40, 50, 65, 80, and 100 nm, which were coated on a Cu substrate to assess the increase in the Rh – B signal at concentration of (10^{-6} M). The results indicated a critical juncture in the correlation between nanoparticle size and maximum SERS intensity. Peak intensity was achieved at a nanoparticle size of 50 nm, subsequently diminishing in accordance with increasing nanoparticle size, as illustrated in **Figure 9** [47].

Conversely, the experiment done by Felix et al. demonstrated that an increase in nanoparticle size correlates with a maximum SERS intensity. Gold nanoparticles of sizes 40, 47, 56, 68, 70, 78, 86, and 98 nm were tested to see how the SERS signal from biphenyl-4-thiol changed and the maximum SERS intensity was at 98 nm, as seen in **Figure 10** [48].

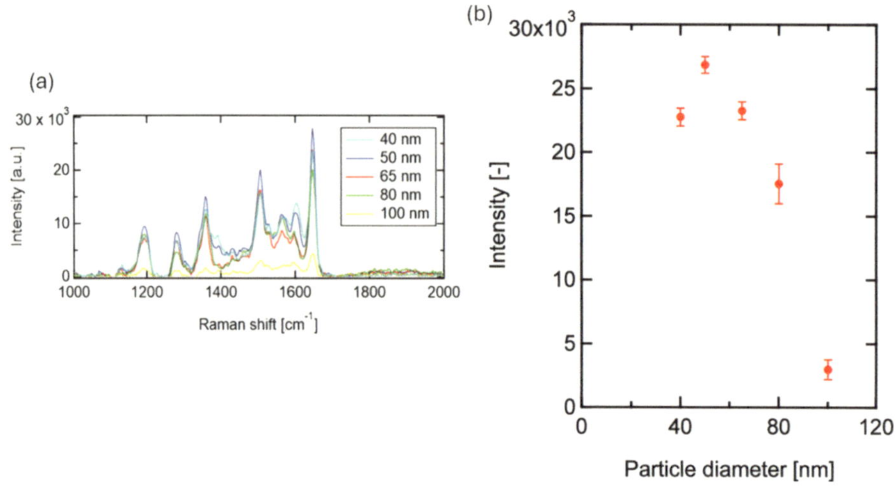

Figure 9.
(a) Outcomes of quantifying the Raman signal's enhancement on nanoparticles of varying size, (b) correlation between nanoparticle size and the intensity of Raman signal enhancement.

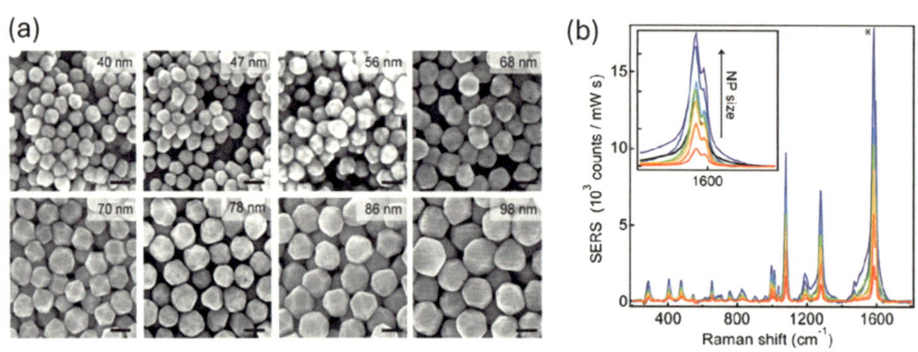

Figure 10.
(a) FE-SEM measurement results, (b) Raman signal enhancement results for different nanoparticle sizes.

3.2.3 Impact of nanoparticle shape

Changing the shape of nanoparticles can also increase Raman sensitivity. Nanoparticles have electrons that will oscillate when exposed to light and produce plasmonics and illustrated as distribution of electric fields. Nanoparticles shaped like triangles and nanostars have electric fields that concentrate at their sharp edges, which means that the plasmonics in these shapes are stronger than in nanorods and round shapes, as demonstrated by the electric field distribution simulation in **Figure 11**.

The shape of silver nanoparticles, whether spherical, triangular, or nanoplate, affects the strength of the methylene blue Raman signal, with triangular nanoparticles giving the strongest signal and spherical ones giving the weakest. The electric field distribution around triangular-shaped nanoparticles is stronger because of their sharp tips compared to the distributions around spherical and nanoplate geometries, as demonstrated in **Figure 12**. When the shape of silver nanoparticles changes, the wavelength of LSPR shifts to longer wavelength. AgNPs-0 is a spherical silver nanoparticle, AgNPs-39 is a triangular nanoparticle, and AgNPs-78 is a nanoplate, as seen in **Figure 12** [52].

The nanostar configuration exhibits the strongest electric field distribution due to its greater number of sharp ends compared to the triangular form. Although the nanorod possesses sharp ends at both extremities, it has fewer sharp ends than the nanostar, resulting in a weaker electric field distribution. Giardo et al. showed this effect by looking at how gold nanoparticles, specifically nanorods and nanostars, optimized oligophenylenedithiol (OPD) Raman signals, as seen in **Figures 12** and **13** [53].

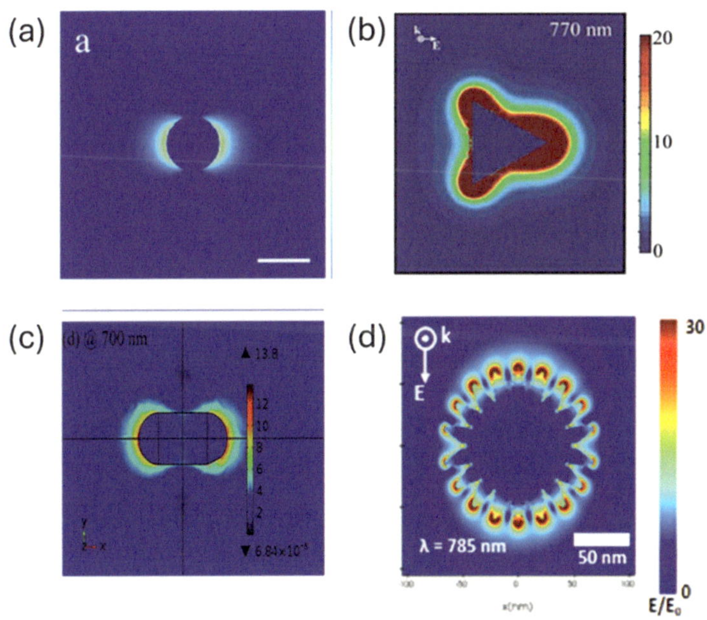

Figure 11.
Electric field distribution on (a) sphere nanoparticles, (b) triangle nanoparticles [49], (c) rod nanoparticles [50], and (d) nanostars [51].

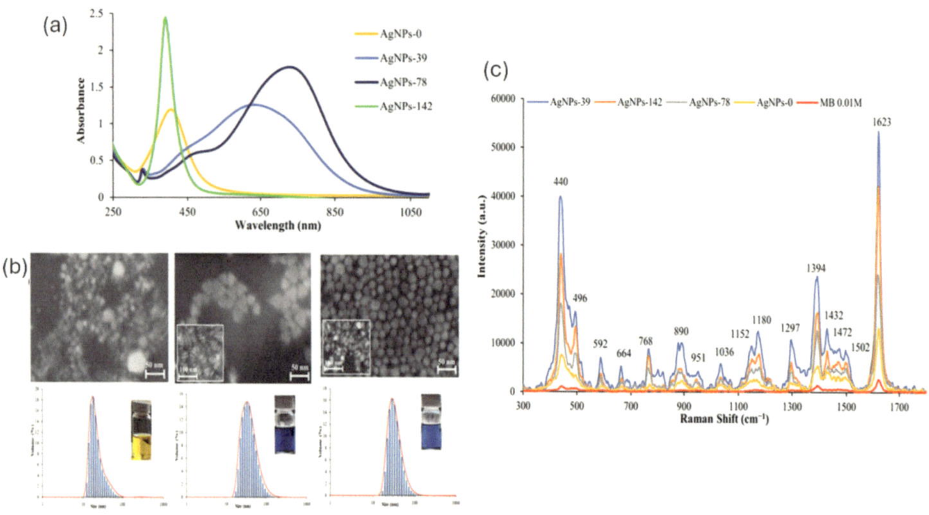

Figure 12.
(a) LSPR of silver nanoparticles spheres, triangles, and nanoplates (b) images of nanoparticles and FE-SEM results and (c) SERS signal enhancement is measured in silver nanoparticles in the shape of spheres, triangles, and nanoplates.

3.2.4 Impact of substrate (surface plasmon polariton)

The enhancement of Raman sensitivity is affected not only by LSPR but also by the choice of substrate, depending regarding the kind of SERS substrate used. In the SERS substrate system, both the localized plasmonic effect of LSPR and implications of plasmon propagation in the substrate and dielectric medium, known as surface plasmon polaritons (SPP), are significant. In the case of solution SERS, only plasmons localized on LSPR nanoparticles, not SPP, impact the Raman signal's enhancement.

EM mechanism on the SERS substrate includes the electromagnetic field emanating from the nanoparticles' plasmonic surface and bouncing off the surface of the substrate material. Thus, it generates a SERS enhancement factor that is significantly influenced by the dielectric properties of the substrate material. This hypothesis is supported by simulating gold nanoparticles shaped like spheres and cubes placed on two different materials: a Si substrate (which is a semiconductor) and a Pt substrate (which is a metal). Using FDTD, the results show that the electric field of Au nanoparticles on the Si substrate is double that of Au nanoparticles on the Pt substrate. The process of merging two varieties of LSPR and SPP plasmons is referred to as plasmon hybridization [22]. SERS enhancement factor average will rise in accordance with the rise in the dielectric material's refractive index. The nanogap between the particle and substrate considerably affects SERS signal amplification, as well as the incident laser's wavelength and polarization states [54]. Xia and his team demonstrated through experiments that plasmon hybridization occurs by increasing the 1,4-BDT signal on Ag nanocubes placed on various materials like Au, Ag, Si, and glass, with the largest signal increase observed on the Ag substrate, followed by Au, Si, and the smallest increase on glass [55].

We conducted analogous experiments in which gold nanoparticles synthesized *via* femtosecond laser were applied to silicon wafer and copper substrates to detect paraquat pesticide. The results showed Raman signal increase. of paraquat on the silicon wafer substrate surpassed that on the copper substrate, likely due to the superior dielectric constant of silicon relative to copper, as illustrated in **Figure 14** [56].

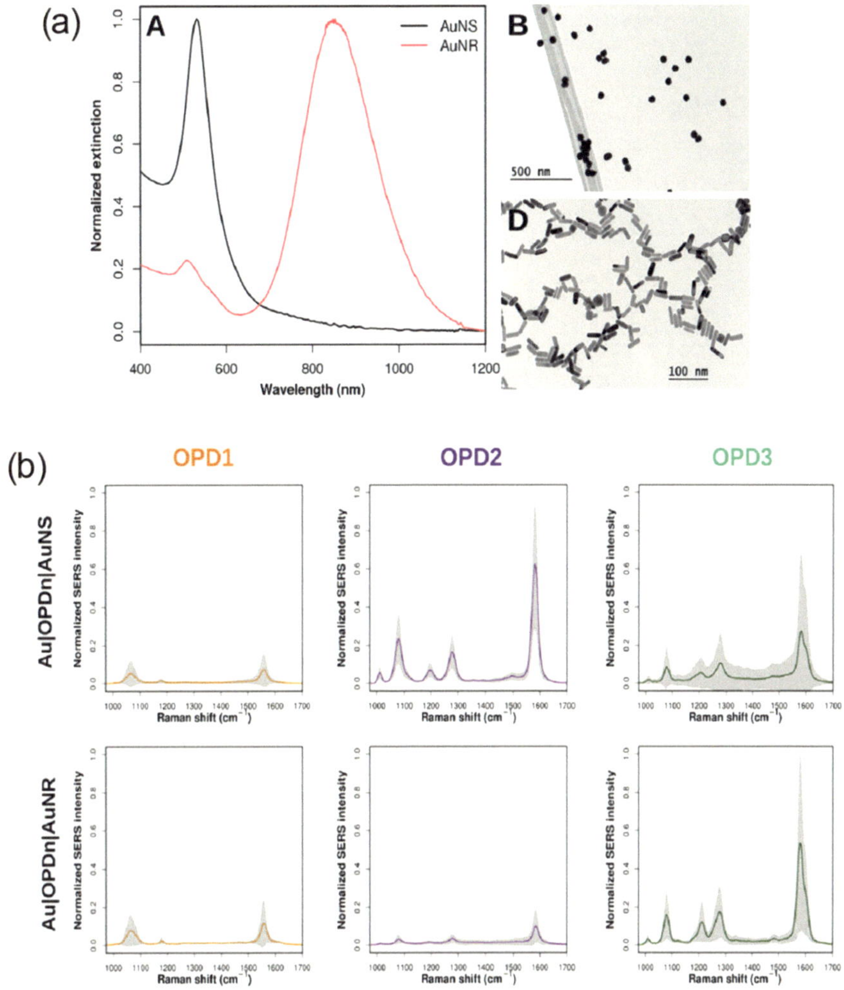

Figure 13.
(a) LSPR of nanostars and nanorods and FE-SEM results and (b) SERS signal enhancement measurements of nanostars and nanorods.

SPP, or surface plasmon polariton, is a longitudinal wave that propagates over surfaces. The surface plasmon polariton (SPP) wave vector far exceeds that of light waves in a dielectric medium. The equation representing the SPP wave vector that propagates along the metal surface is as follows:

$$k_{SPP} = \frac{\omega}{c}\sqrt{\frac{\varepsilon(\omega)\varepsilon_m}{\varepsilon(\omega)+\varepsilon_m}} \tag{3}$$

vacuum light speed is c, ω is the angular frequency, and $\boldsymbol{\varepsilon}(\omega)$ and $\boldsymbol{\varepsilon}_{\mathrm{m}}$ are the metal and dielectric constants, respectively.

Within the metal film layer, the SPP attenuates, reducing the electromagnetic field's intensity. The propagating surface plasmon polariton (SPP) is delineated as follows:

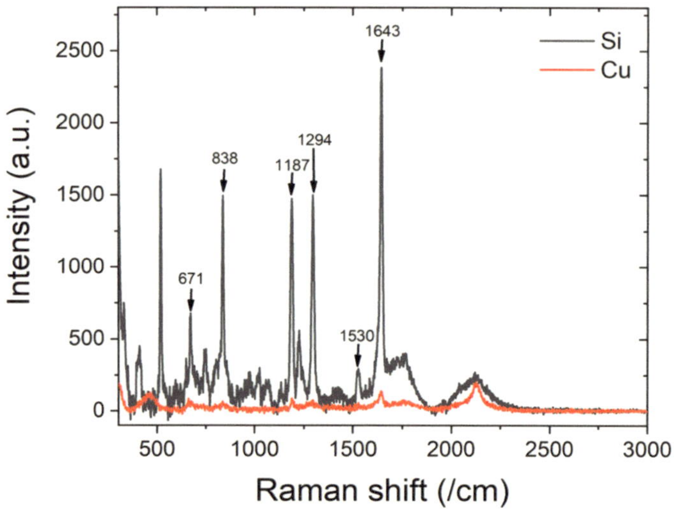

Figure 14.
SERS signal enhancement measurement results of paraquat pesticide on silicon wafer and Cu substrates.

$$L = \frac{1}{2 I_m \left(k_{spp} \right)} \tag{4}$$

$$L_{SPP} \approx \frac{\lambda}{2\pi} \left(\frac{\epsilon'(\omega) + \epsilon_M}{\epsilon'(\omega)\epsilon_M} \right)^{\frac{3}{2}} \frac{\epsilon'(\omega)^2}{\epsilon''(\omega)} \tag{5}$$

where ($\varepsilon = \epsilon' + i\epsilon''$), ($\epsilon_M$) indicates the medium's dielectric constant, and (λ) is the electromagnetic wavelength that strikes the metal layer [57].

4. Conclusions

Unlike traditional methods, the vibrational spectroscopy technique known as Raman spectroscopy can identify harmful analyte contamination more quickly and without causing sample damage. Nevertheless, the sensitivity of Raman spectroscopy for detecting low concentrations is typically inadequate, necessitating the use of nanoparticles with plasmonic characteristics to enhance electromagnetic waves by Raman scattering, hence increasing Raman spectroscopy's sensitivity. The method is Surface Enhanced Raman Spectroscopy.

SERS is categorized into solution SERS and substrate SERS. Liquid SERS only employs a nanoparticle solution that is directly combined with the analyte, resulting in the localization of plasmons around the nanoparticles, specifically localized surface plasmon resonance (LSPR), which enhances the Raman signal. Upon the SERS substrate, the plasmon hybridization phenomenon occurs, integrating LSPR, influenced by the characteristics of the nanoparticles, and surface plasmon polaritons (SPP), which propagate along the substrate layer and are significantly affected by the substrate's properties and dielectric constant. A substrate with a higher dielectric constant, such as a silicon wafer, will yield a more substantial signal enhancement compared to substrates with lower dielectric constants, such as paper and copper. The nanoparticles' shape, size, and kind determine the plasmonic characteristics of LSPR; larger nanoparticles produce smaller hot spots, which strengthens the Raman signal. Nanoparticles shaped like nanostars and triangles create stronger electric fields at their pointed tips, which improves their plasmonic properties compared to round shapes. The nanoparticles shaped like nanostars and triangles would boost the Raman signal, making Raman spectroscopy more sensitive for detecting analyte.

Enhancing the sensitivity of Raman spectroscopy for detecting analytes at low concentrations requires selecting the appropriate substrate, the appropriate nanoparticle type for the analyte, the ideal nanoparticle size for a small hot spot, and the appropriate nanoparticle shape that generates strong plasmons.

Conflict of interest

The authors declare no conflict of interest.

Author details

Affi Nur Hidayah
Research Center for Photonics, National Research and Innovation Agency, Tangerang Selatan, Banten, Indonesia

*Address all correspondence to: affi001@brin.go.id

References

[1] Eberhardt K, Stiebing C, Matthäus C, Schmitt M, Popp J. Advantages and limitations of Raman spectroscopy for molecular diagnostics: An update. Expert Review of Molecular Diagnostics. 2015;**15**(6):773-787. DOI: 10.1586/14737159.2015.1036744

[2] Hashimoto K, Ramaiah Badarla V, Kawai A, Ideguchi T. Complementary vibrational spectroscopy. Nature Communications. 2019;**10**:4411. DOI: 10.1038/s41467-019-12442-9

[3] Paraskevaidi M, Matthew BJ, Holly BJ, Hugh BJ, Thulya CPV, Loren C, et al. Clinical applications of infrared and Raman spectroscopy in the fields of cancer and infectious diseases. Applied Spectroscopy Reviews. 2021;**56**(8-10):804-868. DOI: 10.1080/05704928.2021.1946076

[4] Petry R, Schmitt M, Popp J. Raman spectroscopy–a prospective tool in the life sciences. ChemPhysChem. 2003;**4**:14-30. DOI: 10.1002/cphc.200390004

[5] Yang D, Ying Y. Applications of Raman spectroscopy in agricultural products and food analysis: A review. Applied Spectroscopy Reviews. 2011;**46**:539-560. DOI: 10.1080/05704928.2011.593216

[6] Carter EA, Edwards HGM. Biological applications of Raman spectroscopy. In: Gremlich H-U, Yan B, editors. Infrared and Raman Spectroscopy of Biological Molecules, Practical Spectroscopy Series. Vol. 24. New York, Basel: Marcel Dekker Inc.; 2007. pp. 421-475

[7] de Oliveira Penido CAF, Pacheco MTT, Lednev IK, Silveira L. Raman spectroscopy in forensic analysis: Identification of cocaine and other illegal drugs of abuse. Journal of Raman Specroscopy. 2016;**47**:28-38. DOI: 10.1002/jrs.4864

[8] Ong TT, Blanch EW, Jones OA. Surface enhanced Raman spectroscopy in environmental analysis, monitoring and assessment. Science of the Total Environment. 2020;**720**:137601. DOI: 10.1016/j.scitotenv

[9] Cortizas AM, Lopez-Costas O. Linking structural and compositional changes in archaeological human bone collagen: An FTIR-ATR approach. Scientific Reports. 2020;**10**:17888. DOI: 10.1038/s41598-020-74993-y

[10] Pahlow S, Weber K, Popp J, Wood BR, Kochan K, Rüther A, et al. Application of vibrational spectroscopy and imaging to point-of-care medicine: A review. Applied Spectroscopy. 2018;**72**:52-84. DOI: 10.1177/0003702818791939

[11] Lister AP, Sellors WJ, Howle CR, Mahajan S. Raman scattering techniques for defense and security applications. Analytical Chemistry. 2021;**93**(1):417-429. DOI: 10.1021/acs.analchem.0c04606

[12] Blackie EJ, Le Ru EC, Etchegoin PG. Single-molecule surface-enhanced Raman spectroscopy of nonresonant molecules. Journal of the American Chemical Society. 2009;**131**:14466-14472. DOI: 10.1021/ja905319w

[13] Goel R, Chakraborty S, Awasthi V, Bhardwaj V, Dubey SK. Exploring the various aspects of surface enhanced Raman spectroscopy (SERS) with focus on the recent progress: SERS-active substrate, SERS-instrumentation, SERS-application. Sensors and Actuators, A:

Physical. 2024;**376**:115555. DOI: 10.1016/j.sna.2024.115555

[14] Fleischmann M, Hendra PJ, McQuillan A. Raman spectra of pyridine adsorbed at a silver electrode. Chemical Physics Letters. 1974;**26**:163-166

[15] Jeanmaire DL, Van Duyne RP. Surface Raman spectroelectrochemistry: Part I. Heterocyclic, aromatic, and aliphatic amines adsorbed on the anodized silver electrode. Journal of Electroanalytical Chemistry and Interfacial Electrochemistry. 1977;**84**:1-20

[16] Moskovits M. Enhanced Raman scattering by molecules adsorbed on electrodes—A theoretical model. Solid State Communications. 1979;**32**:59-62

[17] Ozbay E. Plasmonics: Merging photonics and electronics at nanoscale dimensions. Science. 2006;**311**:189. DOI: 10.1126/science.1114849

[18] Raether H. Surface Plasmons on Smooth and Rough Surfaces and on Gratings, Springer Tracts in Modern Physics (STMP). Vol. 111. Berlin, Heidelberg: Springer; 1988

[19] McFarland AD, Young MA, Dieringer JA, Duyne RPV. Wavelength-scanned surface-enhanced Raman excitation spectroscopy. The Journal of Physical Chemistry. B. 2005;**109**:11279. DOI: 10.1021/jp050508u

[20] Pilot R, Signorini R, Durante C, Orian L, Bhamidipati M, Fabris L. A review on surface-enhanced Raman scattering. Biosensors. 2019;**9**:57. DOI: 10.3390/bios9020057

[21] Jing T, Rongke G, Changbiao Z, Yiyue Y, Xiaozhe C, Yang L, et al. Flexible surface-enhanced Raman scattering substrates: A review on design strategies, fabrication technologies, and applications. Coordination Chemistry Reviews. 2025;**539**:216739

[22] Ding SY, Yi J, Li JF, et al. Nanostructure-based plasmon-enhanced Raman spectroscopy for surface analysis of materials. Nature Reviews Materials. 2016;**1**:16021. DOI: 10.1038/natrevmats.2016.21

[23] Smekal A. Zur quantentheorie der dispersion. Die Naturwissenschaften. 1923;**11**:873-875. DOI: 10.1007/BF01576902

[24] Raman CV, Krishnan KSA. New type of secondary radiation. Nature. 1928;**121**:501-502. DOI: 10.1038/121501c0

[25] Landsberg GS, Mandelstam LI. Eine neue Erscheinung bei der Lichtzerstreuung in Krystallen. Die Naturwissenschaften. 1928;**16**:557-558

[26] Jiang Y, Suna D, Pua H, Wei Q. Surface enhanced Raman spectroscopy (SERS): A novel reliable technique for rapid detection of common harmful chemical residues. Trends in Food Science and Technology. 2018;**75**:10-22

[27] Bhavya MB, Rhakho N, Jena SR, Yadav S, Altaee A, Saxena M, et al. Detection of PFAS via surface-enhanced Raman scattering: Challenges and future perspectives. Sustainable Chemistry for the Environment. 2023;**3**:100031

[28] Chimenti RV. Excitation source parameters dictate Raman spectroscopy outcomes. Laser Focus World. 2017. Spectrometers [Internet]. Available from: https://digital.laserfocusworld.com/laserfocusworld/201701?article_id=1252818&pg=NaN#pgNaN [Accessed: June 1, 2025]

[29] Orlando A, Franceschini F, Muscas C, Pidkova S, Bartoli M, Massimo R, et al. A comprehensive review on Raman

spectroscopy applications. Chemosensors. 2021;**9**:262. DOI: 10.3390/chemosensors9090262

[30] Colthup N. Introduction to Infrared and Raman Spectroscopy. New York: Academic; 2012

[31] McCreery RL. Raman Spectroscopy for Chemical Analysis. Chichester: Wiley; 2000

[32] Huang Q, editor. Nanotechnology in the Food, Beverage and Nutraceutical Industries, A Volume in Woodhead Publishing Series in Food Science, Technology and Nutrition. Amsterdam: Elsevier; 2012

[33] Le Ru EC, Etchegoin PG. Principles of Surface-Enhanced Raman Spectroscopy and Related Plasmonic Effects. Amsterdam: Elsevier; 2009

[34] Zamkoye II, Lucas B, Vedraine S. Synergistic effects of localized surface plasmon resonance, surface plasmon polariton, and waveguide plasmonic resonance on the same material: A promising hypothesis to enhance organic solar cell efficiency. Nanomaterials. 2023;**13**:2209

[35] Futamata M. Surface plasmon polariton enhanced Raman scattering from adsorbates on a "smooth" metal surface: The effect of thickness and dielectric properties of constituents. Langmuir. 1995;**11**(10):3894-3901. DOI: 10.1021/la00010a046

[36] Awada C, Dab C, Zhang J, Ruediger A. Observation of propagating surface plasmon polaritons by using functionalized tip-enhanced Raman spectroscopy. Journal of Raman Spectroscopy. 2020;**51**:8. DOI: 10.1002/jrs.5855

[37] Langer J, de Aberasturi DJ, Ja A, et al. Present and future of surface-enhanced Raman scattering. ACS Nano. 2020;**14**:28-117. DOI: 10.1021/acsnano.9b04224

[38] Shan C, Xiaohong L, Yuxiao J, Wei Z, Zhigang Z. Surface enhanced Raman scattering revealed by interfacial charge-transfer transitions. The Innovation. 2020;**1**(3):100051. DOI: 10.1016/j.xinn.2020.100051

[39] Long DA. The Raman Effect a Unified Treatment of the Theory of Raman Scattering by Molecules. Chichester: Wiley; 2002

[40] Jensen L, Aikens CM, Schatz GC. Electronic structure methods for studying surface-enhanced Raman scattering. Chemical Society Reviews. 2008;**37**:1061. DOI: 10.1039/B706023H

[41] Kanipe KN, Chidester PPF, Stucky GD, Meinhart CD, Moskovits M. Properly structured, any metal can produce intense surface enhanced Raman spectra. Journal of Physical Chemistry C. 2017;**121**(26):14269-14273. DOI: 10.1021/acs.jpcc.7b02637

[42] Sharma B, Frontiera RR, Henry AI, Ringe E, Duyne RPV. SERS: Materials, applications, and the future. Materials Today. 2012;**5**:1-2. DOI: 10.1016/S1369-7021(12)70017-2

[43] Hidayah AN, Triyono D, Herbani Y, Saleh R. Liquid surface-enhanced Raman spectroscopy (SERS) sensor-based Au-Ag colloidal nanoparticles for easy and rapid detection of Deltamethrin pesticide in brewed tea. Crystals. 2022;**12**:24. DOI: 10.3390/cryst12010024

[44] Hidayah AN, Triyono D, Herbani Y, Isnaeni SR. The performance of Au-Ag nanoalloys as SERS substrate on silicon wafer in enhancement Raman signal of deltamethrin pesticides. In: AIP Conference Proceedings on 6th

International Symposium on Frontiers of Applied Physics (ISFAP); 6-7 October 2021, Indonesia. Vol. 2652. New York: AIP Publishing; 2022. p. 020001. DOI: 10.1063/5.0107398

[45] Bozo E, Dombovari A, Me M, Virtanen VK, Saarakkala S, Vajtai R, et al. Cu-Pd bimetal and CuPt alloy nanotubes derived From Cu nanowires: Novel amplification media for surface-enhanced Raman spectroscopy. IEEE Sensors Journal. 2020;**20**:1. DOI: 10.1109/JSEN.2019.2939677

[46] Beyene AB, Su WN, Tsai HC, et al. Cu/Ag nanoparticle-based surface-enhanced Raman scattering substrates for label-free bacterial detection. ACS Applied Nano Materials. 2022;**5**:11567-11576. DOI: 10.1021/acsanm.2c02571

[47] Kenmotsu S, Hirasawa M, Tamadate T, Matsumoto C, Osone S, Inomata Y, et al. Surface-enhanced Raman scattering on size-classified silver nanoparticles generated by laser ablation. ACS Omega. 2024;**9**:37716-37723. DOI: 10.1021/acsomega.4c03046

[48] Benz F, Chikkaraddy R, Salmon A, Ohadi H, de Nijs B, Mertens J, et al. SERS of individual nanoparticles on a mirror: Size does matter, but so does shape. Journal of Physical Chemistry Letters. 2016;7:2264-2269. DOI: 10.1021/acs.jpclett.6b00986

[49] Krajczewski J, KołaTaj K, Andrzej K. Plasmonic nanoparticles in chemical analysis. RSC Advances. 2017;**7**:17559-17576. DOI: 10.1039/C7RA01034F

[50] Weiyan L, Jianwen Z, Juan H, Haibibu A, Qiming Z, Hu J. Self-assembly of Au–Ag alloy hollow nanochains for enhanced plasmon-driven surface-enhanced Raman scattering. Nanomaterials. 2022;**12**(8):1244. DOI: 10.3390/nano12081244

[51] Jiamin X, Koen E, Yu-chien L, Taveer AT, Qiyu Z, Ren-Jei C, et al. Near infrared window active magnetic core spiky gold nanostars for dual mode imaging and photothermal therapy. Materials Today Advances. 2025;**25**:100555. DOI: 10.1016/j.mtadv.2024.100555

[52] Zannotti M, Rossi A, Giovannetti R. SERS activity of silver nanosphere, triangular nanoplates, hexagonal nanoplates and quasi-spherical nanoparticles: Effect of shape and morphology. Coatings. 2020;**10**(3):288. DOI: 10.3390/coatings10030288

[53] Paganoto GT, Temperini MLA, dos Santos DP. Nanoparticle shape effects on the SERS properties of small conductive bridges. Journal of Physical Chemistry C. 2024;**128**:15974-15984. DOI: 10.1021/acs.jpcc.4c03587

[54] Ding SY, You EM, Tian ZQ, Moskovits M. Electromagnetic theories of surface-enhanced Raman spectroscopy. Chemical Society Reviews. 2017;**46**:4042. DOI: 10.1039/C7CS00238F

[55] Rycenga M, Xia X, Moran CH, Zhou F, Qin D, Li ZY, et al. Generation of hot spots with silver nanocubes for single-molecule detection by surface-enhanced Raman scattering. Angewandte Chemie, International Edition. 2011;**50**:5473-5477

[56] Hidayah AN, Triyono D, Herbani Y, Saleh R. Effect of Cu and Si wafer substrates in increasing Raman signal of surface-enhanced Raman scattering-based Au nanoparticles. Malaysian Journal of Science. 2025;**44**(2):78-85. DOI: 10.22452/mjs.vol44no2.8

[57] Yitian L, Yaoguang M. One-dimensional plasmonic sensors. Frontiers in Physics. 2020;**8**:Article 312. DOI: 10.3389/fphy.2020.00312

Section 3

Applications

Chapter 5

Application of Raman Spectroscopy in the Research of Materials for Hydrogen Storage

Olena Zavorotynska

Abstract

Storage and transportation of hydrogen is one of the fundamental sectors of the hydrogen economy value chain. Currently, large quantities of hydrogen are stored either at low temperature (20–250 K) or high pressure (up to 75 MPa), which requires costly technological tanks and poses considerable safety issues. Storing hydrogen in materials has been proposed as an alternative solution utilizing material matrices that bind either molecular or atomic/ionic hydrogen. In this field, the studies of hydrogen bonding and dynamics are central. Raman spectroscopy (RS) has been used as a powerful tool for characterizing molecular vibrations to understand the structural and chemical properties of hydrogen storage materials. This chapter presents the basic aspects of Raman scattering relevant to the field of hydrogen storage materials and reviews recent advances in applications. The main aspects discussed are vibrational properties of the H_2 molecule and the effect of adsorption, Raman crystallography with respect to studies of metal hydrides, utilization of Raman scattering to investigate hydrogen dynamics, *in situ* and *operando* Raman studies of hydrogen sorption in porous materials, hydrogen diffusion, and hydrogenation/dehydrogenation reactions in metal hydrides.

Keywords: hydrogen storage, hydrogen adsorption, surface characterization, hydrogen absorption, hydrides, Raman crystallography, isotope effect

1. Introduction

Hydrogen is an energy storage medium that is seen as a key player in green economies and global energy security due to its high energy content per mass unit (140 MJ kg^{-1}). This content is three times that of gasoline and natural gas and almost two orders of magnitude larger than that of the most advanced Li-battery (711 Wh / kg or 2.56 MJ kg^{-1} as of 2023 [1]). However, the low density of H_2 gas results in its drastically low volumetric energy content that poses significant challenges for storage, transportation, and utilization. For practical applications, hydrogen gas is currently compressed, with the typical storage pressures between 35 and 70 MPa, liquified at 20 K, or cryocompressed, reaching up to 80 kg m^{-3} hydrogen densities (**Figure 1**). The extreme operational temperatures and pressures bring up the cost of the technology and pose significant safety risks [7].

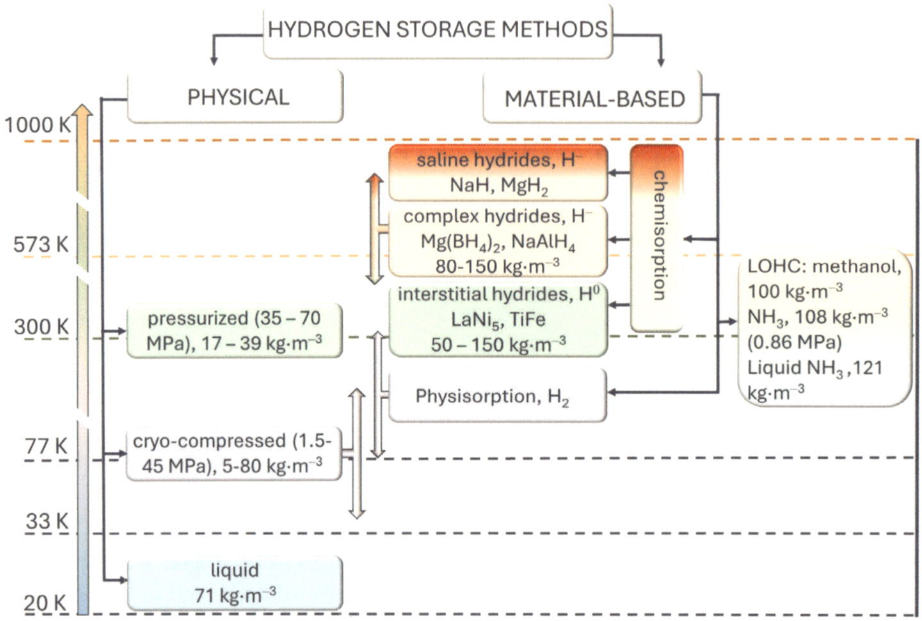

Figure 1.
Classification of hydrogen storage methods based on operational temperature (storage, desorption, ad–/ absorption). Hydrogen volumetric density is indicated for each method if available. The figure does not consider activation temperature conditions for interstitial hydrides (ca. 650 K) and materials for physisorption (ca. 400K). Refs. [2–6].

An alternative method of densifying hydrogen is material-based storage. In this method, hydrogen is stored either in molecular form *via* adsorption on a surface of a high-surface-area matrix or as a part of a chemical compound as H, H^+, or H^-. Storage of adsorbed H_2 still requires cryogenic temperatures, but those are usually above 77 K, which is the temperature of liquid N_2, a widely available cryogenic cooler. Adsorption can also be employed at near ambient temperature, utilizing non-ambient pressures. In either case, higher gas densities can be achieved at significantly improved operational T and p compared to those of the compressed or liquified hydrogen (**Figure 1**), bringing down the technology cost and safety risks [7, 8].

The storage matrix potentially should adsorb or absorb and release hydrogen fast and safely at practical T and p. Among other required properties are light weight, low cost, non-toxicity, and availability of raw materials. The adsorbents for molecular hydrogen storage must possess a high specific surface area (SSA) for accommodating large gas quantities. The values of about 1000 $m^2\,g^{-1}$ and above are of practical interest. Optimal porosity and strong adsorption sites for fast diffusion kinetics and favorable thermodynamics (e.g., T, p operation conditions) are required [9–11].

Alternatively, ionic or atomic hydrogen can be stored in hydrides. Various classes of hydrides exist with drastically distinct chemical and hydrogen storage properties, as described in Section 3.1.1. They incorporate metal (metal hydrides) or nonmetal elements (NH_3, NH_3BH_3, organic compounds) that reduce gravimetric hydrogen density (usually defined in wt% H_2) compared to those of pure compressed or liquid H_2. The volumetric hydrogen densities, however, in these compounds can greatly suppress those of liquid H_2, reaching almost double, 150 Kg H_2 per m^3 of the compound in Mg_2FeH_6 (note: at NTP!) [12]. Metal hydrides can store 1–21 wt% H_2 at an operational T-range between ambient and 850 K (**Figure 1**). The release of hydrogen typically occurs at near ambient pressure, but rehydrogenation requires elevated temperatures and pressures of H_2. Chemical hydrides, such as NH_3BH_4 or NH_4BH_4, have high hydrogen content by weight and volume but typically are irreversible [3]. Ammonia gas and liquid and liquid organic carriers (LOHCs), such as methanol and organosilanes [13], are yet another way to store and transport hydrogen.

Optimization of material hosts for hydrogen storage has mainly been focused on affecting the interaction with molecular or ionic/atomic hydrogen, catalytic splitting of H_2 in the case of chemisorption, and improved diffusion. Naturally, Raman spectroscopy (RS), as an excellent tool to probe molecular bonding and lattice vibrations, is one of the valuable techniques in this research, characterizing both adsorbed H_2 and chemically bonded hydrogen. The method is relatively inexpensive and accessible in many research laboratories. Raman scattering can be observed from gas, liquid, crystalline, and amorphous solids, making the technique highly universal. Owing to instrumental advances and relatively fast data collection, *in situ* and *operando* Raman measurements enable real-time observation of hydrogen adsorption/absorption and desorption. Compared to another vibrational spectroscopy technique, Fourier-transformed infrared spectroscopy (FTIR), RS provides complementary information about molecular vibrations. However, RS has several advantages: it can access a wider analysis energy range in one instrument (ca. 100–5000 cm^{-1}), whereas FTIR typically requires two distinct setups: mid-IR and far-IR. RS does not require sample dilution in an optically transparent matrix, such as transmission and diffuse-reflectance IR, and thus has a higher versatility for *in situ* studies in various sample environments.

Neutron spectroscopy, in particular inelastic neutron scattering (INS), is another technique to study structure and dynamics in hydrides due to an exceptionally high hydrogen-scattering cross-section for neutrons, deep penetration through certain metals, absence of selection rules, and straightforwardness of theoretical spectra calculation [14, 15]. However, INS requires much larger sample quantities, isotope-substituted compounds in some cases (D, ^{7}Li, ^{11}B), and neutron measurements are only available in a few large-scale laboratory facilities with restricted proposal-based access. Neutron experiments are thus very expensive and require well-in-advance planning.

This chapter summarizes fundamentals and recent progress and highlights the potential of utilizing Raman spectroscopy in the research of hydrogen storage materials. The chapter is aimed at early-stage researchers who wish to extend the method in their work beyond routine phase identification. In the first part, the fundamentals of Raman scattering from H_2 gas, liquid, and adsorbate are described with examples from the literature. The second part is focused on Raman spectroscopy in the solid state, explaining basic features of Raman scattering from crystalline compounds with the representative case studies.

Because LOHCs and polysilanes are organic compounds and Raman studies on them are adjacent to the application of Raman spectroscopy in organic chemistry, this class of hydrogen storage materials will not be considered here. The chapter assumes basic knowledge of the molecular vibrations theory, the classical and quantum-mechanical foundations of the Raman scattering phenomenon, the polarizability tensor, and the Raman spectrometer; thus, these topics will not be detailed here.

2. Raman spectroscopy of molecular hydrogen

2.1 Hydrogen isotopes and molecular nuclear isomers

Hydrogen atom exists in three natural isotope forms with one electron, one proton, and zero, one, or two neutrons in the nucleus: protium, H (natural occurrence 99.985%); deuterium, D (0.015%); and tritium, T (10^{-18}%); respectively [16]. The three isotopes have different atomic masses and nuclear spins (1/2 for H and T, and 1 for D). All the hydrogen isotopes form diatomic molecules: H_2, D_2, T_2, HD, DT, and HT. For H_2, the H–H bond length (proton-proton separation) for the minimum interaction potential energy is 0.74611 Å with a binding energy of 4.52 eV, or 218.1 kJ mol^{-1} H per H atom [16].

Molecular H_2, D_2, and T_2 are homonuclear diatomic molecules, with the consequence that special symmetry restrictions are placed upon their wavefunctions. This results in two nuclear isomers: ortho—the species with the largest spin degeneracy—and para—the species with the lowest spin degeneracy. Consequently, there are two kinds of hydrogen molecules for H_2: orthohydrogen (o-H_2, symmetric, with parallel nuclear spins ↑↑) and parahydrogen (p-H_2, antisymmetric, with antiparallel nuclear spins ↑↓) [17].

In addition to o-H_2, p-H_2, equilibrium hydrogen (e-H_2), and normal hydrogen (n-H_2), are defined. E-H_2 (e-D_2 and e-T_2) all correspond to the equilibrium concentration at given temperatures. N-H_2 is a mixture of the equilibrium concentration of o-H_2 and p-H_2 at a temperature of 293.15 K (20°C), which is 25 mol% of p-H_2 and 75 mol% of o-H_2. This ratio preserves upon liquefaction [17]. However, equilibrium liquid and solid hydrogen are constituted mostly (up to 100%) by p-H_2. Ortho-para conversion is forbidden for isolated molecules, but it occurs in cryogenic gas and liquid/solid, driven by the magnetic dipole interaction of the nuclear moments of o-H_2 molecules. The conversion rate of o-H_2 to p-H_2 is T-dependent and very low (years at 77 K) but can be catalyzed with surface-active or paramagnetic species to occur in minutes [2]. Upon conversion, the heat is released, ΔH_{conv}, which increases with lowering temperature from 270 kJ kg^{-1} at 300 K to 523 kJ kg^{-1} below 77 K. Ortho-para conversion is an important process that must be taken into consideration in liquefaction and cryogenic adsorption, as the energy released during the conversion is higher than the heat of evaporation, causing boil-off effect and hydrogen loss during storage [2].

2.2 RS of molecular hydrogen at ambient and non-ambient conditions

Molecular hydrogen has been an important test ground in the early development of quantum mechanics and is still a recognized benchmark for tests of quantum electrodynamics and of theories that describe physics beyond the standard model [18, 19]. Therefore, the spectra of H_2 and the isotopes have been meticulously studied and measured with a few parts-per-billion uncertainty [19]. For our purposes and in the conditions of a standard Raman experiment, the values with few or no decimal figures will suffice.

The Raman spectrum of hydrogen gas recorded in a typical Stokes range (100–4600 cm^{-1}) is composed of lines corresponding to the molecular transitions between vibrational, rotational, and roto-vibrational levels in the molecule, which are distinct for ortho- and para-isomers.

In calculating the vibrational transitions, it is important to include anharmonicity effects (Eq. (1)), as the potential of light diatomic molecules deviates significantly from the harmonic oscillator approximation [20, 21]:

$$E_n = hc\omega_e\left(n + \frac{1}{2}\right) - hcx_e\omega_e\left(n + \frac{1}{2}\right)^2 + \ldots \tag{1}$$

Where ω_e is the wavenumber corrected for the anharmonicity, and $x_e\omega_e$ indicates the magnitude of anharmonicity: $x_e\omega_e$ (H_2) = 117.91 cm^{-1}—the largest for all diatomic molecules—and $x_e\omega_e$ (D_2) = 64.10 cm^{-1} [20]. The selection rules for the quantum harmonic oscillator define $\Delta n = \pm 1$, but the large anharmonicity allows for the forbidden vibrational transitions ($\Delta n > |1|$), and they can be observed in the experimental spectra with low intensity.

Hydrogen isotopes demonstrate the largest *isotope effect,* defined as the shift in vibrational frequencies in the spectra. For H_2 and D_2, the shift can be calculated according to Eq. 2 [20]:

$$\rho = \frac{\nu_{H2}}{\nu_{D2}} = \sqrt{\frac{\mu_{(D2)}}{\mu_{(H2)}}} = 1.41 \tag{2}$$

where μ – is the reduced mass, and ν – is the respective vibrational mode.

The nuclear spin is associated with the rotational motion of the nucleus described by rotational number J. When the nuclei are fermions (non-integer spin: H, T), the symmetric nuclear states must be coupled with the antisymmetric rotational states (odd J), and the antisymmetric nuclear spin states must be coupled with the symmetric rotational states (even J) [17, 21]. Considering the nuclear spin isomers, the selection rule for the rotational transitions for the hydrogen molecule is $\Delta J = 0, \pm 2$, defined by the selection rules for rigid rotor transitions. Even though real diatomic molecules are not strictly rigid rotors, they do not deviate from the selection rule $\Delta J = 0, \pm 2$, as in the case of the vibrational levels of anharmonic quantum oscillator [21]. The allowed rotational transitions for H_2 and D_2 are summarized in **Table 1** and are much lower in energy for D_2 due to the larger mass and moment of inertia.

Hydrogen preserves almost free rotation even at the extremely low temperatures in the solid state, where $J = 0$ are predominantly populated levels in equilibrium [17].

The energy region for pure rotational spectral lines (S type) is found in 300–900 cm^{-1}, vibrational in 4100–4220 cm^{-1} (Q type), and roto-vibrational in the 4750–4450 cm^{-1} regions (**Table 2**). For example, the $S_0(0)$ and $S_0(2)$ peaks at 354 cm^{-1} and 814 cm^{-1} are due to the $J_{0\rightarrow 2}$ and $J_{2\rightarrow 4}$ rotational transitions of p-H_2, respectively.

	H_2		D_2	
J	**Exp.**	**Theory**	**Exp.**	**Theory**
0	354.38	354.39	179.06	179.12
1	587.06	587.07	297.52	297.63
2	814.41	814.48	414.66	414.78
3	1034.65	1034.75	529.91	530.07

Table 1.
Experimental and theoretical rotational quanta for H2 and D2 molecules in the ground vibrational state (in cm^{-1}) [17].

The $S_0(1)$ peak at 587 cm^{-1} is due to the rotational transitions $J_{1 \rightarrow 3}$ of o-H_2. The intensity ratio reflects the relative natural abundance of the para and ortho spin isomers. Thus, at room temperature, the peak intensity ratio of para-to-ortho H_2 is 1:3 [22].

The Q-type peaks at 4161, 4155, 4143, and 4126 cm^{-1} are all due to pure H-H vibrational transitions satisfying the $\Delta n = 1$ and $\Delta J = 0$ selection rules: the peaks at 4161, $Q_1(0)$, and 4143 cm^{-1}, $Q_1(2)$, originate from para-H_2 molecules occupying the $J = 0$ and the $J = 2$ rotational levels of the ground vibrational state, respectively; those at 4155, $Q_1(1)$, and 4126 cm^{-1}, $Q_1(3)$, by ortho-H_2 molecules in the $J = 1$ and $J = 3$ states. Upon condensation, i.e., in the liquid and solid state, due to the intermolecular interactions, the S_0 rotational lines show broadening of a few cm^{-1}. Q-transitions demonstrate a red shift by 7–9 cm^{-1} in the liquid and by 8–11 cm^{-1} in the solid from the gas values [21, 23, 24].

RS was demonstrated to accurately assess hydrogen isomer composition in industrial processes related to hydrogen liquefaction. A straightforward and physically rigorous single-point calibration approach utilizing o-H_2/p-H_2 ratios of Raman intensities of the rotational modes 200–1400cm^{-1} with the error in concentration measurements below 1.09% between T = 77–343 K was developed [25].

Change in shape and position of the Raman lines has been used to study phase transitions (PhT) and discover new phases in ultra-compressed H_2 (above 250 GPa, at 300 K) [26], contributing to the studies of the H_2 phase diagram [27]. In more practical conditions, it was shown that the position of Q-branches was affected only within at most 1 cm^{-1} at up to 400 °C and 40 MPa. At the same time, the position, intensity, and peak width of the molecular transitions are much affected during dissolution in water and seawater due to hydrogen molecular interactions and can be developed into a direct approach to determine the concentration of dissolved H_2 in high-temperature hydrothermal fluids [28].

2.3 Raman studies of physisorbed H_2

Molecular H_2 reversibly adsorbs on surfaces generally due to van der Waals interactions. The spectroscopic studies of H_2 adsorption in high-surface-area porous materials can reveal the magnitude and nature of the interactions with the surface between the adsorbate molecules and provide such characteristics of the surfaces as relative surface area and the nature of the adsorption sites. As the adsorption is a surface phenomenon, a nearly linear relation between surface area and adsorption capacity exists, where the matrices with higher surface area exhibit higher H_2 uptakes [29].

	H_2			D_2		
	gas	**liquid**	**solid**	**gas**	**liquid**	**solid**
Q1(0)	4161	4154	4152	2993	2987	2985
Q1(1)	4155	4146	4143	2991	2984	2982
Q1(2)	4143					
Q1(3)	4126					
S1(0)	4500					
S1(1)	4715					

Table 2.
Vibrational, $Q1(J)$, and rotational-vibrational, $S1(J)$, frequencies (in cm^{-1}) of H2 and D2 in the free state (gas, liquid, and solid phases) [21, 23, 24].

The interaction of H_2 with surfaces is generally very weak and can usually be monitored at low T or high p. The H–H stretching has been shown to undergo a red shift ($\Delta\nu$(H-H)) with respect to free H_2. In fact, the adsorption process polarizes H_2 such that it acquires a dipole moment and becomes active in FTIR, which has led to vast research in the field of *in situ* variable temperature infrared spectroscopy (VTIR) for H_2 adsorption [11, 30]. In addition, as the single H-H bond is weak, the effect of even rather similar adsorption sites can be distinguished by slight differences in the $\Delta\nu$(H-H), and the step-by-step formation of various adducts can be followed [31, 32].

H_2 interaction energy during physisorption is typically measured by the isosteric method that presumes a constant value throughout the whole adsorption [9, 33]. This isosteric heat of adsorption, Q_{ist}, is a mean value of the interaction energy. H_2 forms extremely weak van der Waals bonds, resulting in isosteric heats of adsorption that are, for carbon-based materials, typically lower than 7 kJ mol^{-1}. Theoretically predicted optimal adsorption enthalpy was shown to be 1.5 and 100 bar at 298 K of 13.6 kJ mol^{-1} over the entire H_2 uptake curve, or even 22–25 kJ mol^{-1} taking into account the enthalpy-entropy correlation [34]. This range is above the weak van der Waals interactions, and therefore, strong adsorption sites should be introduced to improve the adsorption and storage conditions. Alternatively, a typical material with an average adsorption enthalpy of 6 kJ mol^{-1} would function optimally at a temperature of 131 K [34, 35]. It was shown, however, that in the case of transition metal hydrazides, hydrides, [36] and organometallic copper species supported on carbon [37], which adsorb H_2 through the Kubas interaction, adsorption enthalpy increases with coverage. Q_{ist} should thus be distinguished from site-specific adsorption enthalpy, ΔH_{ads}. ΔH_{ads} characterizes the energy of interaction with a particular adsorption site. To provide the ΔH_{ads} constant over all ranges of adsorption, the pores of the storage material should be small enough to accommodate 1–3 layers of adsorbed H_2 molecule [38], in particular around 10 Å for a slit cylindrical pore shape [39]. Increasing the H_2 binding energy within the surface of a storage material is the most important challenge for creating hydrogen adsorbents that operate at near ambient conditions.

2.3.1 Characterization of surface adsorption sites

H_2 storage *via* physisorption has been explored in many different types of high-surface-area materials. Most typical adsorption sites, in order of increasing adsorption energy, include:

i. organic linkers in organic polymers, covalent organic frameworks (COFs), and metal-organic frameworks (MOFs): $\sim$ 4–6 kJ mol^{-1} [10, 22, 40],

ii. zeolitic framework: bridged -OH groups, coordinatively unsaturated framework sites, surface silanols (-Si-OH), and partial negative charge on zeolite framework: 5–18 kJ mol^{-1} [11, 41],

iii. coordinatively unsaturated edges and corners in high-surface-area nanosized oxides: 3–7 kJ mol^{-1} [31],

iv. electrostatic interaction with coordinatively unsaturated metal sites: 3–17.5 kJ mol^{-1} [10, 32],

v. Kubas complexes with framework 3d transition metals (TMs): 20–80 kJ mol^{-1} [36, 42].

In the latter Kubas complexes with transition metals, H_2 donates its bonding electron pair onto the partially filled 3d orbital of the metal and can accept electrons on its antibonding orbital σ^* (backdonation), resulting in a side-on (η^2) bonding in M-η^2-H_2 complex where H_2 acts as a weak Lewis base. The bonding is between strong physisorption and weak chemisorption [42]. It can be affected by the type of metal and presence of other ligands, thus opening the vast possibilities for material engineering in various types of metal-decorated frameworks. Kubas complexes induce elongation of the H-H bond from 0.8–1.0 Å to 1.6 Å, after which it breaks, forming a hydride. The bond elongation results in the significant shift of ν(H-H) that can then be found in the 2082–3205 cm^{-1} range, and in addition, five fundamental modes: $\nu_{as}(MH_2)$ (1345–1575 cm^{-1}), $\nu_s(MH_2)$ (850–950 cm^{-1}); two deformations, $\delta(MH_2)_{in\text{-}plane}$ and $\delta(MH_2)_{out\text{-}of\text{-}plane}$ (450–563 cm^{-1}); and a torsion (H_2 rotation), $\tau(H_2)$ [42].

RS under 150 bar of H_2 was employed to observe Kubas in vanadium alkyl hydrides: the peaks at 2770, 2929, and 3834 cm^{-1} were all assigned to ν(HH) of H_2 complexes with three different absorption types, including V(III) and V(IV) [36], whereas in V(III) polymers, one peak at 2553 cm^{-1} was observed [43], demonstrating the effect of cation environment on interaction energy. Similar complexes were observed in Cr alkyl hydrides with the peaks at 2789, 2922, and 3188 cm^{-1} assigned to the complexes as confirmed with D_2 studies [44].

In the case of alkali and alkaline-earth, the interaction is governed essentially by polarization contributions and is strongest for small divalent metal cations that are also most coordinatively unsaturated and accessible for guest H_2 molecules. The interaction induces a dipole moment in H_2 that makes FTIR studies possible. In fact, the vast majority of spectroscopic characterization of H_2 adsorption in porous materials has been performed with the FTIR/VTIR method. It has been noted that a roughly linear shift exists between the H_2-Me interaction energy, ΔH_{ads}, and the red shift of hydrogen vibrational frequency, $\Delta\nu$(H-H), although the exceptions are not rare. The shift can be up to 150 cm^{-1} but is most typically within 60–100 cm^{-1} [10, 11, 30, 31, 34, 40].

2.3.2 Vibrational response of material matrix upon gas sorption

Changes in the Raman signature of the adsorption framework can also be used as an indirect indication of gas adsorption, and in the case of H_2, especially when molecule dissociation is involved [11, 30]. Modification of framework modes in the 1000–1600 cm^{-1} region of HKUST-1 MOF functionalized with carbon nanotubes was observed during H_2 adsorption at RT [45]. Pd/MoO_3 has shown the disappearance of MoO_3 modes upon interaction with H_2, which was explained by crystal transformation from original α-MoO_3 into H_xMoO_3 in the presence of H_2 [46].

3. Raman spectroscopy of chemically stored hydrogen

3.1 Hydrides for hydrogen storage: H, H-, H+

Hydrogen exhibits a variety of bonding when forming compounds with metals and nonmetals, adopting considerably different sizes in different oxidation states (**Figure 2**). Metal-hydrogen compounds can thus be divided into several classes with very different hydrogen storage properties: thermodynamic stability (i.e., operational conditions), hydrogen content (**Figure 1**), hydrogen absorption/desorption reaction

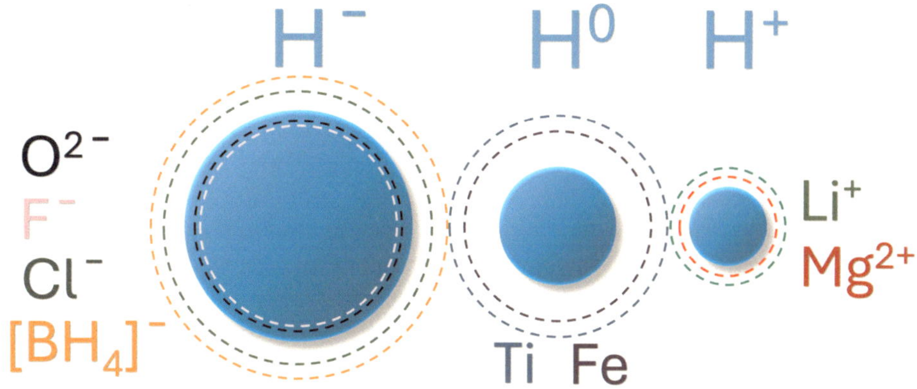

Figure 2.
Oxidation states of hydrogen and atomic radii in comparison to some other relevant elements/ions.

pathways, and reaction kinetics [3, 14]. Transition metal alloys, such as TiFe and $LaNi_5$, multicomponent alloys, or medium/high entropy alloys (HEAs), and some transition metals (i.e., Pd) form *interstitial hydrides*. Interstitial hydrides split the H_2 molecule and store atomic H in the voids within the metal lattice. These hydrides typically exhibit metallic bonding and retain the metal's electrical conductivity. Many interstitial hydrides can absorb, store, and release H_2 at near ambient temperatures and practical pressures with fast kinetics of adsorption/release that make them of great practical interest for material-based hydrogen storage. The metallic nature of these compounds, however, does not allow for extensive Raman studies, as Raman scattering is hindered by screening of the charge by the longitudinal electric field [47]. For this reason, Raman studies of such compounds are scarce.

On the opposite side of the thermodynamic stability scale are ionic, or *saline, hydrides*. These are the hydrides of alkali and alkaline-earth metals where anionic hydrogen (H^-) is incorporated in a typically cubic lattice in alkali and orthorhombic for alkaline-earth environments, except for MgH_2 that crystallizes in a tetragonal lattice with the space group $P4_2/mnm$. Bonding in saline hydrides is predominantly ionic. This affects the decomposition temperature of the hydrides, which ranges from 170°C (RbH, CsH) to 720°C (LiH) [5]. Due to the light weight of Group I and II metals, the gravimetric hydrogen content in saline hydrides is relatively high, ranging between 0.75 and 18.16 wt%. The volumetric hydrogen density within 26–138 kg m^{-3} [5] is also practically interesting (the higher values). For hydrogen storage applications, MgH_2 has received by far the most attention due to the combination of satisfactory hydrogen content (7.6 wt% and 110.0 kg m^{-3}), relatively low decomposition temperature (327°C), availability of nontoxic Mg, and straightforward decomposition/rehydrogenation reaction with good kinetics of H_2 release and absorption [3, 48]. CaH_2, NaH, LiH, and KH, although they do not satisfy the requirements for hydrogen content alone and have high decomposition temperatures, have been studied as part of *reactive hydride composites* (RHC) and as thermal energy storage media, where the high stability is an advantage [49]. RHC are the mixtures of several compounds with high hydrogen content and catalytic additives, for example, LiH/MgB_2 [50, 51], CaH_2/MgB_2 [52], $Mg(BH_4)_2/MeH_x$ (Me = Li, Na, Mg, Ca) [53], $2LiNH_2/MgH_2/MeH$ (Me = K, Rb, Cs) [54], etc. The mixing is intended to increase the overall hydrogen storage capacity and/or improve the kinetics and thermodynamics of the whole system through self-catalyzing reactions.

Hydrogen content of Group I and II metal-based compounds can be increased if H^- is substituted with another hydrogen-rich anion such as $(BH_4)^-$, $(AlH_x)^{y-}$, or $(NH_x)^{y-}$, $[SiH6]^2$-creating *complex hydrides*: borohydrides, alanates, amines, amides and imides, and silanides (silicides) [55–57], respectively. These compounds form ionic solids with molecular hydrogen-rich anions. The gravimetric hydrogen content in the practically interesting complex hydrides (Li, Na, K, Mg, and Ca-based) varies within 5.71–18.4 wt%, which is their main advantage, and the volumetric content is 70.3–124.1 H kg m^{-3} [5]. The disadvantages of these compounds are high thermodynamic stability, complex decomposition/rehydrogenation pathways with slow kinetics, sensitivity to moisture and oxygen that requires handling under an inert atmosphere, and poor reversibility of hydrogen absorption. Hydrogen desorption in complex hydrides occurs above 120°C and often is limited by the formation of the saline hydride, which is again thermodynamically stable. Cycling of hydrogen, often partially irreversible, requires higher temperatures and pressure above 10 MPa [3, 58, 59].

In *complex transition metal (TM) hydrides,* the hydrogen-rich anion is formed by hydrogen covalently bonding with a central transition metal: $[NiH_4]^{4-}$, $[FeH_6]^{4-}$, $[CoH_5]^{4-}$, $[RuH_6]^{4-}$, etc. The anions then are imbedded in the lattice of alkali, alkaline-earth, or rare-earth cations, forming the compounds of the type $M_x^{y+}[TH_z]^{y-}$: Mg_2NiH_4, Mg_2FeH_6, or also mixed-cation compounds such as $CaMgNiH_4$. Due to the ability of TM to possess a different number of oxidation and coordination states, there exists a large variety of TM complex hydrides, most of them interesting from a fundamental, structural point of view, and due to potential applications in catalysis, smart windows, thermal storage, and high-temperature superconductivity [60, 61]. Only a few of them, composed of lighter elements, such as those mentioned above, have been considered for hydrogen storage. Mg_2FeH_6 stores 5.6 wt% hydrogen, and its volumetric hydrogen density is the highest of all known compounds: 150 kg m^{-3} [12]. In addition, TM complex hydrides have more straightforward H_2 release pathways without producing toxic gases, and improved reversibility. However, their formation and decomposition temperatures above 200°C are still too high for practical hydrogen storage applications [60–63].

There has been a vast amount of research in recent years directed at optimizing hydrides for hydrogen storage. The research effort is mainly focused on destabilizing the covalent X-H bonds in the molecules/molecular cations or promoting the bond formation for improved reversibility by adding catalysts [64–66], creating self-catalyzing RHC, anion substitution with a halogen [67–70], or nanosizing through dispersion on a high-surface-area matrix or mechanical milling [65].

Raman spectroscopy is an excellent tool to probe bonding in hydrogen-containing molecules, molecular ions, and lattice H– of the solid-state compounds. **Figure 3** provides an overview of the spectral regions where the scattering of X-H bonds can be observed, which can be used as a quick reference for the identification of hydrogen bonds in the compounds. Due to the sensitivity of Raman scattering to the molecular symmetry, it can detect subtle shifts in vibrational frequencies that indicate phase transitions, bonding environments, and the presence of intermediate species, providing valuable insights into the structure of the hydrides and mechanisms of hydrogen release and absorption reactions. Additionally, isotopic substitution of H with D can provide isolation of the anions of interest, assisting in the studies of reaction pathways and hydrogen diffusion. In the following, we will present an overview of the fundamentals of Raman scattering in solid state with a focus on hydrides, with illustrative examples of such studies.

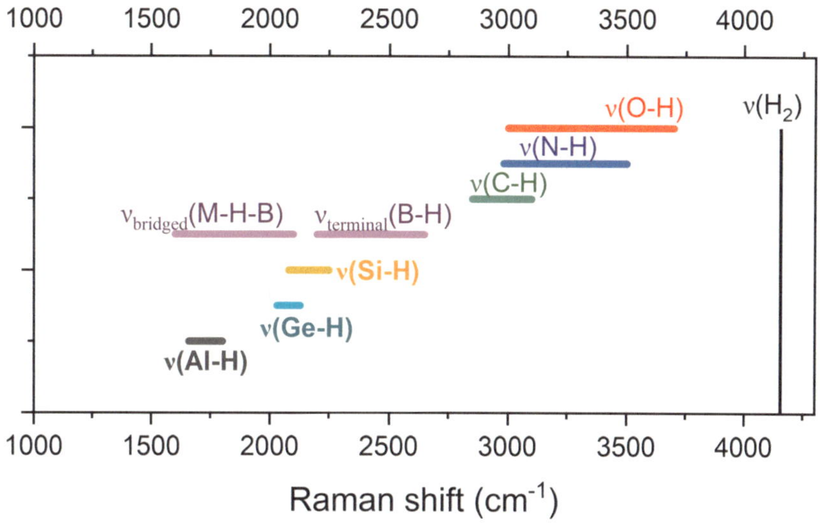

Figure 3.
X-H stretching frequency with elements. Ref. [20, 56, 71, 72].

3.2 Structural studies of hydrogen storage compounds with Raman scattering

3.2.1 Molecular symmetry and selection rules

Based on its structure, each molecule can be classified into a symmetry point group: C groups (C_p, C_{ph}, C_{pv}) have one symmetry axis of order p and a plane either perpendicular (σ_h) or through it (σ_v) (symmetry elements of each group are defined by the subscript of the group). D groups (D_p, D_{ph}, D_{pd}), in addition to C_p, have pC_2 symmetry axes perpendicular to it and at equal angles to one another, and can have D_{ph}—a perpendicular to C_p plane σ_h and $p\sigma_v$ planes through C_p at angles of 360°/2p to one another. Group D_{pd} has $p\sigma_d$, which goes through the C_p and bisects the angles between two successive C_2 axes. S_p group has a p-fold roto-reflection axis S_p (p is even). The T_d group has three mutually perpendicular C_2 axes, four C_3 axes, and a σ_d through each pair of C_3 axes. Many hydrogen-rich ions adopt T_d symmetry (**Table 3**). O_h molecules have three mutually perpendicular C_4 axes, four C_3 axes, and a center of symmetry, i. Finally, I_h group has 6 C_5 axes, 10 C_3 axes, 15 C_2 axes, 15 σ planes, and a center of symmetry I, summing up to 120 symmetry elements. This symmetry is observed in the B_{12} skeleton of $B_{12}H_{21}^{2-}$ or fullerene C_{60} [20]. Each point group has a set of defined normal modes of a given symmetry as summarized in *character tables* that can be found in any textbook on vibrational spectroscopy, for example, Appendix 1 in ref. [20]. A *character*, χ, of the vibration is +1 for the symmetric and − 1 for the antisymmetric modes.

Doubly (E) and triply (F or T) energy degenerate vibrations are found in molecules with an axis higher than C_2 and C_3, respectively. Other common notations of the normal mode symmetries are A and B to denote nondegenerate modes, that are symmetric with respect to respect to rotation about the principal axis (z), and asymmetric, respectively. If two modes in the same point group differ in the character of C, other than the principal axis, they are further distinguished by subscripts 1, 2, and 3. If two modes differ in the character of σ other than σ_v, they are distinguished by ′ and ″ [20].

Molecule: point symmetry	Examples	Normal modes (symmetry)	Assignment/activity in IR, Raman	Ref.
XH_2: C_{2v}	$[NH_2]^-$	$\nu_1(A_1)$ $\nu_2\ (A_1)$ $\nu_3\ (B_2)$	ν_s (X-H_2) / IR, R δ_s (H-X-H) / IR, R ν_{as} (X-H)/ IR, R	[20]
XH_3: C_{3v}	NH_3, SiH_3	$\nu_1(A_1)$ $\nu_2\ (A_1)$ $\nu_3\ (E)$ $\nu_4\ (E)$	ν_s(X-H_3) / IR, R δ_s(H-X-H) / IR, R ν_d(X-H) / IR, R δ_d(H-X-H) / IR, R	[73]
XH_3: D_{3h}	BH_3, AlH_3, GaH_3	$\nu_1(A'_1)$ $\nu_2\ (A''_2)$ $\nu_3\ (E')$ $\nu_4\ (E')$	ν_s(X-H_3) / IR, R π (X-H_3) / IR ν_d(X-H) / IR, R δ_d(H-X-H) / IR, R	[20]
XH_4: T_d	$[BH_4]^-$, $[GaH_4]^-$, $[AlH_4]^-$, $[NH_4]^+$ CH_4, SiH_4, GeH_4,	$\nu_1(A_g)$ $\nu_2\ (E)$ $\nu_3\ (F_2)$ $\nu_4\ (F_2)$	ν_s(X-H_4) / R δ_s(H-X-H) / R ν_{as}(X-H) / IR, R δ_{as}(H-X-H) / IR, R	[20]
XH_5: C_{4v}	$[CoH_5]^{3-}$	$\nu_1(A_1)$ $\nu_2\ (A_1)$ $\nu_3\ (A_1)$ $\nu_4\ (B_1)$ $\nu_5(B_1)$ $\nu_6\ (B_2)$ $\nu_7\ (E)$ $\nu_8\ (E)$ $\nu_9\ (E)$	$\nu(X-H)_{top}$/IR, R ν_s(X-H_4) / IR, R π (X-H_4) / IR, R ν_{as}(X-H_4) / R δ_{as}(X-H_4) / R δ_s(X-H_4) / R ν_d(X-H_4) / IR, R π_d (X-H_4) / IR, R δ_d(X-H_4) / IR, R	[20, 61]
XH_6: O_h	$[AlH_6]^{3-}$ $[FeH_6]^{4-}$ $[SiH_6]^{2-}$	$\nu_1(A_{1g})$ $\nu_2\ (E_g)$ $\nu_3\ (F_{1u})$ $\nu_4\ (F_{1u})$ $\nu_5\ (F_{2g})$ $\nu_6\ (F_{2u})$	$\nu\ (X-H)_{in-phase}$ / R $\nu\ (X-H)_{out-of-phase}$ / R ν (X-H) / IR δ(H-X-H) / IR δ(H-X-H) / R δ(H-X-H) / n.a.	[20, 56, 74]

π – out-of-plane symmetric bending.

Table 3.
Molecular symmetry and normal modes of the most common hydrogen-containing ions and molecules.

The number of normal vibrations of each symmetry can be calculated based on the equations given in Appendix III of ref. [20].

To determine the activity of the vibrations in the infrared and Raman spectra, the selection rule (change in at least one component of the polarizability tensor) must be applied to each of them. The selection rules have been derived quantum mechanically and applied to individual molecules using group theory. This information is readily available in literature (Ref. [20]). Using symmetry properties and character tables, the selection rule for Raman-active modes can be rephrased as the vibration is Raman-active if it belongs to the same symmetry species as one of the components of the polarizability tensor.

3.2.2 Symmetry and normal modes of hydrogen-containing molecules and molecular ions

Hydrogen-rich ions adopt various symmetries that affect the number and Raman activity of normal modes, as is summarized in **Table 3**. For example, ammonia, NH_3, is a pyramidal molecule of the point symmetry C_{3v}. All its four normal modes are Raman-active. In this type of molecule, an inversion doubling can be observed, where two configurations of NH_3 are equally probable, which is manifested as the splitting of its ν_1 and $\nu_2$modes into two [73]. XH_3 neutral molecules of BH_3, AlH_3, and GaH_3 have planar D_{3h} geometry with only three Raman-active modes: ν_1, ν_3, ν_4, in contrast to the pyramidal XH_3 molecules. In XH_4 tetrahedral molecules, trends $\nu_3 > \nu_1$ and $\nu_2 > \nu_4$ hold for most of the compounds [20].

Boron forms an enormous kingdom of compounds with hydrogen, and the chemistry of boranes is a subject for another textbook. **Table 3** lists only vibrations of the tetrahedral $[BH_4]^-$. Spectroscopic analysis of others, such as important intermediates in the decomposition of borohydrides (i.e., $Mg(B_3H_8)_2$, $B_{10}H_{10}$ (D_{4d}), $B_{12}H_{12}$ (I_h)), and B_2H_6 as a byproduct of the dehydrogenation reactions, can be found in dedicated reviews and reference textbooks [20, 75]. Parker et al. reviewed spectroscopic properties of complex TM hydrides (compounds of $[PdH_2]^{2-}$, $[PdH_3]^{3-}$, $[PdH_4]^{2-}$, $[PdH_4]^{4-}$, $[PtH_4]^{4-}$, $[PtH_5]^{3-}$ $[RuH_6]^{4-}$, $[PtH_6]^{4-}$, $[ReH_9]^{2-}$) that are beyond the scope of hydrogen storage applications and this work [61].

3.2.3 Effects of isotope substitution and ion exchange on molecular symmetry

Isotope H → D substitution causes substantial changes in the vibrational spectra due to the symmetry lowering and new X-D modes that are clearly separated in the spectra because of the drastic difference in masses of *H* and *D* [76]. Symmetry lowering causes splitting of degenerate vibrations according to the rules summarized in correlation tables. Correlation tables indicate all possible subgroups that can be obtained from a point group when its symmetry is lowered. An example of the normal modes splitting upon H → D substitution for the T_d point group of XH_4 anions is shown in **Table 4**.

The Redlich-Teller product rule [73] (eq. 3) relates the vibrational frequencies ν_n and ν'_n of pure and isotopically substituted molecules. The rule is written as:

$$\frac{\nu_1\nu_2\nu_3\ldots\nu_n}{\nu'_1\nu'_2\nu'_3\ldots\nu'_n} = \sqrt{\left(\frac{m'_1}{m_1}\right)^{\alpha}\left(\frac{m'_2}{m_2}\right)^{\beta}\cdots\left(\frac{M}{M'}\right)^{t}\left(\frac{I_x}{I'_x}\right)^{\delta_x}\left(\frac{I_y}{I'_y}\right)^{\delta_y}\left(\frac{I_z}{I'_z}\right)^{\delta_z}} \tag{3}$$

We will explain the equation and its application using substitution in $[BH_4]^-$ (the brackets and negative sign will be omitted in the following for brevity) as an example, in which $\nu_n = \nu_4$. The $m_1, m_2, \ldots$ are masses of the representative atoms of the various sets of equivalent nuclei, which in the case of BH_4 and BD_4 are H (m_1 = 1 amu), D (m_1' = 2 amu), and B ($m_2 = m_2'$ = 10.81 amu [1]). The α and β are the coefficients of m_1 and m_2 in the equations calculating the number of normal modes in each symmetry species [20, 73]. The relative equations are summarized in **Table 5**.

BX_4 ions of the T_d symmetry have only the nuclei n_3 (number of sets of nuclei on a threefold axis: H(D)) and n_0 (number of sets of nuclei on all symmetry elements, which is B in the center of the tetrahedron). N_3 in the equations of **Table 5** corresponds to m_1 in eq. 3, and similarly n_0 corresponds to m_2 in eq. 3). This reasoning can be verified by the fact that BX_4 tetrahedral ions have one A_1, one E, and two F_2 modes

---

[1] Natural boron is constituted by 80% of ^{11}B and 20% of ^{10}B. Here the weighted average of this composition is adopted for briefness.

Point group	Normal modes of vibrations and their assignment			
T_d (XH_4)	**ν_1 (A_1)**	**ν_2 (E)**	**ν_3 (F_2)**	**ν_4 (F_2)**
C_{3v} (XH_3Y)	ν_1 (A_1) ν_s of XH_3	ν_5 (E) δ_d of XH_3	ν_2 (A_1) ν of XY	ν_3 (A_1) δ_s of XH_3
			ν_4 (E) ν_d of XH_3	ν_6 (E) ρ_r of XH_3
C_{2v} (XH_2Y_2)	ν_1 (A_1) ν_s of XH_2	ν_3 (A_1) δ_s of XH_2	ν_2 (A_1) ν_s of XY_2	ν_4 (A_1) δ_s of XY_2
		ν_5 (A_2) ρ_t of XH_2	ν_6 (B_1) ν_{as} of XH_2	ν_7 (B_1) ρ_w of XH_2
			ν_8 (B_2) ν_{as} of XY_2	ν_9 (B_2) ρ_r of XH_2
C_{3v} (XY_3H)	ν_1 (A_1) ν_s of XY_3	ν_5 (E) δ_d of XY_3	ν_2 (A_1) ν of XH	ν_3 (A_1) δ_s of XY_3
			ν_4 (E) ν_d of XY_3	ν_6 (E) ρ_r of XY_3
T_d (XY_4)	ν_1 (A_1) ν_s of XY_4	ν_2 (E) δ_s of XY_4	ν_3 (F_2) ν_{as} of XY_4	ν_4 (F_2) δ_{as} of XY_4

ν_s and ν_a – symmetric and antisymmetric stretching; δ_s, δ_a, and δ_d – symmetric, antisymmetric, and degenerate bending; ρ_t, ρ_w, and ρ_r – twisting, wagging, and rocking deformations, respectively.

Table 4.
Normal mode transformations for the (XH4-xYx) ions with Td, C3v, and C2v symmetries, derived after examples in p. 83–84 and p. 204 in Ref. [20] All modes are Raman-active.

Symmetry	Number of vibrations	Values used in eq. 8	Ratio ν/ν'
T_d A_1 (ν_1)	$3n + 2nd + n_2 + n_3 = 1$	$n_3 = 1$ (X), $\alpha = 1$; $\beta = t = \delta_{x,y,z} = 0$	1.414
E (ν_2)	$6n + 3nd + n_2 + n_3 = 1$	$n_3 = 1$ (X), $\alpha = 1$; $\beta = t = \delta_{x,y,z} = 0$	1.414
F_2 (ν_3, ν_4)	$9n + 5nd + 3n_2 + + 2n_3 + + n_{0\_}1 = 2$	$n_3 = 1$ (X), $n_0 = 1$ (B); $\alpha = 2$ ($2n_3$); $\beta = 1$ ($1n_0$), $t = 3$; $\delta_{x,y,z} = 0$	1.397

Table 5.
Formulas for calculating the number of normal vibrations for each symmetry species for the BX_4 (point group T_d) and the calculated ratios between the frequencies of vibrations of each symmetry.

(**Table 5**). The equation for F_2 modes in **Table 5** gives the coefficients $\alpha = 2$ and $\beta = 1$. M and M' in Eq. 3 are the total masses of pure and substituted molecules (14.81 for BH_4 and 18.81 for BD_4), and t is the number of T_x, T_y, and T_z translations of the given symmetry. For F_2, $t = 3$, and it is zero for A_1 and E. The $\delta_x, \delta_y, \delta_z$ define whether rotational modes R_x, R_y, and R_z belong to the symmetry considered. In T_d groups, they belong to F_1 symmetry. Thus, δ_x, δ_y, $\delta_z = 0$ for all the modes. In this way, $\nu_{3,4}(BH_4)$ / $\nu_{3,4}(BD_4)$ is calculated as 1.397. For ν_1 (A_1) and ν_2 (E), $\alpha = 1$, and all other power coefficients are zero, and the isotopic ratio is $\sqrt{2} = 1.414$. This value is typically used to estimate the isotopic shifts in the deuterated molecules. In this way, H → D substitution in BH_4 produces B-D stretching modes in the 1600–1700 region, which is confirmed experimentally [77].

Upon H → D substitution, the species with different degrees of substitution coexist, as was shown for partially deuterated $LiBH_4$, all $(BH_{4-x}D_x)^-$ $(0 < x < 4)$, where the complex Raman spectra of all the present isotopic species are deconvoluted and assigned [78].

Similar effects of the symmetry lowering can be expected due to anionic substitution into an XH_y group (in which case eq. 3 is obviously not applicable).

3.2.4 Isotope exchange and anion substitution studies in complex hydrides

Isotope exchange is an essential method in spectroscopy for assigning normal modes, as isotopically labeled bonds have well-predicted vibrational shifts. In addition to this application, in complex hydrides it has been a focus of several studies aiming at the fundamental insights into the phenomena relevant to hydrogen release and absorption, such as breaking and formation of B-H bonds [65, 77–80], hydrogen transport and diffusion [78, 81], reaction mechanisms [82] (infrared study) [83], and verification of isotopic purity for neutron-diffraction studies [84, 85].

The conditions for H → D substitution can be used as an indication of the strength of the X-H bond. For example, deuteration of the alkaline (Li, Na, K) borohydrides occurs in the 200–500°C T-range, at several tens of bars of D_2 and requires many hours [80, 86, 87]. In $Ca(BH_4)_2$ and α-$Mg(BH_4)_2$ (dense phase of $Mg(BH_4)_2$), the breaking of a B-H bond and substitution occurred at significantly lower temperatures, D_2 pressures, and shorter reaction times than those for the alkaline borohydrides. The activation energies (E_a) of the deuterium-hydrogen exchange reactions in $Ca(BH_4)_2$ (ca. 80 kJ/mol) [80] and α-$Mg(BH_4)_2$ (ca. 50 kJ/mol) [79]. In the porous high-surface-area γ-polymorph of $Mg(BH_4)_2$, the H → D substitution onset was recorded at even milder conditions, i.e., several minutes at ~80–100°C and only 3 bar D_2 [77]. Notably, the comparable amounts of deuterium substituted in the denser α-$Mg(BH_4)_2$ polymorph required 23 hours under ~40 bar at 132°C [79]. This observation was partially enabled by the rare porous structure of the γ-phase of $Mg(BH_4)_2$ and its high specific surface area (SSA $\approx 1000\ m^2 \cdot g^{-1}$) [88], allowing for the investigations *in situ* by means of Raman spectroscopy, similarly to the *in situ* spectroscopic studies of gas adsorption in high-surface-area solids. The deuteration conditions were even milder in the γ-$Mg(BH_4)_2$ milled with additives, indicating the positive effect of additives and/or milling on weakening of the B-H bond [65].

Raman studies of H-D substituted RHC elucidate reaction pathways and diffusion processes. Bulk and surface exchange of hydrogen in a $LiBH_4$ + $LiNH_2$ mechanochemically prepared mixture containing an intermediate phase $Li_2BH_4NH_2$ was studied with isotope labeling. Surface exchange processes occurred prior to decomposition and with the appearance of the Li_2NH phase with high ionic conductivity. Bulk transport was determined as the rate-limiting step for the H_2 desorption reaction [89]. The study of the synthesis mechanism of $LiBH_4$ in LiD + B_2H_6 reaction revealed that the entire BH_4^- unit is transferred from B_2H_6 to the Li^+ cation, providing clear evidence for the heterolytic splitting of diborane on alkali hydrides and implying the exchange of BH_4^- with H^- ions of the underlying hydride [90]. High mobility of H was concluded for $LiAlH_4$ when the mixed ion compounds appeared after milling the $LiAlH_4$/$LiAlD_4$ mixture, even at room temperature, whereas no easy exchange was observed in the ionically substituted $Li_4(BH_4)(NH_2)_3$ even in the melt [91].

The effects of symmetry lowering can be expected due to anionic substitution into an XH_y group. In particular, in an attempt to destabilize/stabilize complex hydrides, the substitution of an XH_y with a halide anion has been proposed [67, 92–99]. It was predicted that while larger halide anions would substitute a whole XH_y group, F^- can substitute one or more hydrogen atoms inside the group (see **Figure 2**). This would make a very interesting spectroscopic case with the appearance of X-F and mixed H-X-F peaks in the Raman spectra [71].

Although the combination of Raman spectroscopy and isotope/ion substitution holds significant potential, the limited number of such studies suggests that this approach remains underutilized in the investigation of hydrogen storage materials.

3.2.5 Symmetry lowering in solid-state: Site symmetry and crystal field splitting

When a molecular ion is imbedded in a crystalline lattice, its symmetry is usually lowered due to the interaction with the surrounding crystal field—*site symmetry effect*. Site symmetry is the symmetry position of the center of gravity of the group, usually a central atom X in XH_y anions. Alterations in the molecular symmetry can lead to splitting of degenerate modes and activation of the previously inactive modes based on the group theory rules of the group tables (example in **Table 4**). Changes in the ion geometry induce frequency shifts relative to those observed in the gas phase. Further, there are typically several molecular ions in a unit cell, determined by the number Z in

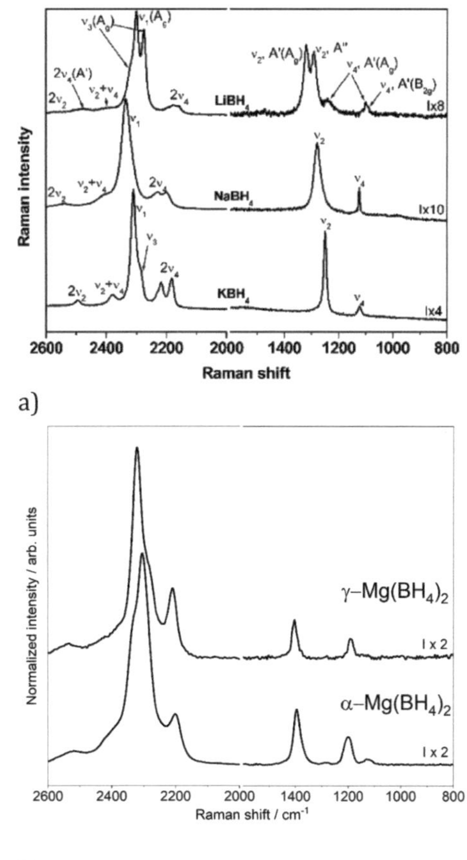

Figure 4.
Experimental Raman spectra of selected alkaline borohydrides and Mg(BH4)2: (a) reprinted with permission from the publisher from ref. [71]; (b) from the author's collection, obtained with a 532 nm laser.

the crystallographic information file (cif). The normal modes are then classified within the symmetry operations of the whole unit cell, and their total number is equal to 3 N-3, where the excluded 3 modes are the translational motions of the whole unit cell—*acoustic phonons*.

The adopted symmetry of the ion in a crystal must be (i) a subgroup of the molecular symmetry of the free ion and (ii) a subgroup of the space group of the crystal [20]. This again can be found in the correlation tables. If the structure of the compound is known, the site symmetry of the anion (usually the central element that sits at the center of mass) can be read in the .cif file.

This interplay between spectroscopy and crystallography can be well illustrated with the example of borohydrides (**Figure 4**). Thus, upon symmetry lowering, BH_4 molecular ions with original T_d symmetry can adopt T, D_{2d}, C_{3v}, S_4, D_2, C_{2v}, C_3, C_2, and C_s site symmetries in a crystalline lattice of a borohydride [20]. In combination with different Z, the set of Raman-active vibrations for each unit cell (*irreducible representation* of Raman-active modes, Γ_R) also differs. The entire spectrum in the 1000–2600 cm^{-1} region is composed of vibrations of BH_4; however, those spectra exhibit distinct variations in the crystalline lattices of different borohydrides due to site symmetry and crystal field effects. In isostructural cubic lattices of alkaline borohydrides (Na, K, Rb, Cs), the anion preserves high symmetry, and the spectra of these compounds are relatively simple, demonstrating degenerate E and F_2 modes. Whereas upon symmetry lowering, these modes split in $LiBH_4$ and in $Mg(BH_4)_2$ (C_s site symmetry but $P6_133$ space group in α-$Mg(BH_4)_2$ and Id-3a in γ-$Mg(BH_4)_2$) (**Figure 4**).

^{10}B / ^{11}B isotope effect. Due to 20% of ^{10}B in natural boron, the isotope effect can also be distinguished in some spectra. For the B-H bond, calculated with the Redlich-Teller product rule (Eq. 3), the difference of 3–15 cm^{-1} in B-H stretching and bending vibrations can be expected. In experimental spectra at ambient temperature, this can manifest in the peak broadening or, in some compounds, such as KBH_4, $RbBH_4$, and $CsBH_4$, in the appearance of a shoulder at a higher frequency of about ¼ the intensity of the main peak [100].

Anion and cation substitution in the structure of hydrides, as well as the formation of solid solutions, can thus be diagnosed by Raman spectra [101].

3.2.6 Lattice vibrations of complex hydrides and vibrations of lattice hydrogen

Lattice vibrations in complex hydrides include rotational and translational modes of the molecular ions and translational modes of the counterions. These typically occur at low frequencies (< 500 cm^{-1} for the rotations and < 450 cm^{-1} for the translations), but their overtones can appear in the region of intermolecular vibrations of the complex anions [61, 71].

In saline hydrides, Raman spectra originate from the translational motion of H^- and counteraction. Lattice dynamics in LiH ($Fm\overline{m}3m$) is well studied both experimentally and theoretically, as it has only four electrons in the unit cell and constitutes the simplest ionic solid [102]. Due to the selection rules for a rock-salt structure (the translation modes have F_{1u} symmetry), the first-order transitions in LiH are not Raman-active. The same is valid for NaH, KH, RbH, and CsH with the $Fm\overline{m}3m$ structure. But the crystal of LiH exhibits very strong and rich second-order Raman scattering ($F_{1u} \times F_{1u} = A_{1g} + E_g + F_{1g} + F_{2g}$, all Raman-active overtones) where 11 peaks can be resolved up to 2500 cm^{-1} [102]. The peaks at 1662, 1814, and 2087 cm^{-1} were assigned to H^- overtone translations, and those at 631, 731, 786, and 825 cm^{-1} to Li^+ motions. The peaks at 486, 1121, 1221, and 1309 cm^{-1} were explained by the coupling of Li^+ and H^- motions. For the absence of the first-order Raman spectra of the alkali hydrides, it is difficult to find measurements in the literature. The lattice dynamics, however, were assessed theoretically [103] and with neutron scattering [104].

MgH_2 crystallizes in a tetragonal lattice ($P4_2/mnm$) with five Raman-active modes per unit cell ($A_{1g} + B_{1g} + B_{2g} + E_g$), and the spectrum exhibits four peaks (B_{2g} mode is very weak) where the E_g and A_{1g} phonons display the strongest scattering efficiency [105]. All these modes, at 262 (B_{1g}), 954 (E_g), and 1280 cm^{-1} (A_{1g}), are due to H^- motion, although those below 320 cm^{-1} are mixed with Mg modes. In addition, there are weaker scattering features at 400–800 cm^{-1} and 1050–1520 cm^{-1} assigned to overtones [105]. CaH_2, SrH_2, and BaH_2 crystallize in an orthorhombic lattice (*Pnma*) with six Raman-active modes ($2A_g + B_{1g} + 2B_{2g} + B_{3g}$). The Raman spectrum of CaH_2 shows 4 peaks at 180, 239, 290, and ca. 840 (very broad) cm^{-1} [106], out of which the 840 peak(s) can be assigned to H^- motion by comparison to INS data [107].

Although Raman spectroscopy of metallic hydrides is not common, interstitial atomic hydrogen was observed in $LaNi_{3.4}Co_{1.2}Mn0.1Al_{0.3}$ metallic hydride in 1 MPa H_2 by using a high detection sensitivity scattering system [108]. Broad peaks at 1600 and 2900 cm^{-1} were assigned to the lattice modes of interstitial atomic H. The authors studied the effect of surface polishing on the Raman spectra. It was concluded that characteristic Raman spectra rarely appear for the polished flat surface and that cracks in the surface play an important role in the hydride formation.

Thus, to accurately analyze spectra of crystals, it is essential to perform a site group and factor group analysis of the whole Bravais cell, which in turn can assist in structure determination. For example, Carsteanu et al. [109], with the help of factor group analysis of YH_x / YD_x and Raman spectra, determined an intermediate phase *I*4/mmm between x = 2 and x = 3. This phase was theoretically predicted and detected by neutron-diffraction experiments. The crystal structure of YH_3 was similarly corrected when factor group analysis indicated that the Raman spectra were not compatible with the proposed $P\bar{3}c1$ but rather with $P6_3cm$ or $P6_3$ space group [110].

There exist excellent resources that perform vibrational analysis of crystals and suggest irreducible representations of crystalline compounds with the analysis of Raman and infrared activity of normal modes based on the .cif input. As an example, Bilbao Crystallographic Server can be recommended [111]. Raman and hyper-Raman scattering applications [112] provide full vibrational analysis of the lattices of interest.

3.2.7 Crystallography with Raman: Polarized Raman spectra on single crystals

The full potential of the Raman experiment can be exploited by performing polarized Raman measurements on single crystals at cryogenic temperatures in combination with factor group analysis and theoretical calculations for the spectra assignment [113]. Low temperature allows for minimizing thermal broadening and enables the highest possible resolution of peaks in the spectra, as well as observation of isotope effects. A polarized Raman beam allows for selective activation of normal modes and, in combination with theoretically predicted positions of Raman shifts, allows for unambiguous spectra assignment. Examples of such studies are on $LiBH_4$ [114] and LiH [102] single crystals at 4–10 K in combination with factor group analysis of normal modes, comparison against theoretical spectra, and/or isotope substitution that allowed for the most reliable spectroscopic analysis.

3.2.8 Resonances and anharmonicities in Raman spectra of hydrides: Overtones, combination bands, and Fermi resonances

Often, a Raman spectrum exhibits more peaks than the predicted number of normal modes. This is due to the fact that combination modes (e.g., $\nu_3 + \nu_2$) and overtones (e.g., $2\nu_2$) can activate due to the anharmonicity of normal vibrations. Symmetry and Raman activity of combinations and overtones are predicted by multiplication rules for irreducible representations (summarized in Appendix IV of ref. [113]). The intensity of these peaks is typically much lower than that of the normal modes unless it is enhanced by Fermi resonance (FR). FR has been observed in borohydrides where the bending modes H-B-H are located at about half the energy of the stretching modes. FR causes mixing of overtones and for a successful modeling of the spectra of such compounds, a variational approach beyond the harmonic approximation supported by computations of B3LYP-type anharmonic force fields is required [115].

3.3 *In situ* Raman and multi-technique studies

Due to the versatility of the Raman experimental setup, various in situ measurements can be carried out in single-technique and multi-technique modes. The laser spot supplied by a remote optical probe can be focused through many types of transparent to visible light sample holders (glass, quartz, halide optical windows, diamond and sapphire windows, etc.). The multi-technique studies allow probing the system's properties at different times, dimensions, etc., scales, and/or assert cross-validation of the results. For example, by combining Raman measurements with diffraction techniques (X-ray, neutron, electron), both long-range and short-range order in the compounds can be characterized, as well as the components in solid/crystalline, amorphous, liquid, and gaseous states.

3.3.1 Phase transitions

As site symmetry and crystal field effects can drastically change the Raman fingerprint of a molecular Raman, the technique is well-suited to study phase transitions (PhT). PhT and change in $[BH_4]^-$ symmetry/Raman spectra of $LiBH_4$ were used for temperature calibration of an *operando* Raman cell [116]. For calibration, the *in situ* Raman data of $LiBH_4$ were measured at RT - 350°C to observe the orthorhombic-$LiBH_4$ → hexagonal-$LiBH_4$ transition between 111°C and the melting point at 275°C. PhT from α to β in $Mg(BH_4)_2$ were studied *in situ* from 15 to 473 K by combined Raman and neutron scattering [117]. The Raman spectra revealed evident signatures of the α → β phase transition. LT spectra allowed for better spectral assignment due to the removal of thermal broadening and increased peak resolution. *Ama*2 (74° C) → *Pnma* (102°C) → $Fm\bar{3}m$ PhT in CsB_3H_8 were for the first time observed by combined *in situ* Raman and synchrotron powder X-ray diffraction (SR-PXRD) measurements [118]. The two high-temperature phases, namely β-CsB_3H_8 and γ-CsB_3H_8, could only be observed at the elevated temperatures, as CsB_3H_8 converts back to α-CsB_3H_8 (*Ama*2) upon cooling.

Pressure-induced PhT can be studied using a high-pressure diamond anvil cell (DAC). Thus, a reversible transition in CaH_2 was found in the 15.5–21 GPa pressure region [119], in $NaNH_2$ at around 1 and 2 GPa [120], and in $NaAlH_4$ above 14 GPa [121].

3.3.2 Reaction mechanisms of hydrogen desorption and adsorption

Decomposition and rehydrogenation of MgH_2 in Ar and 100 bar H_2, respectively, were monitored *in situ* by Raman *via* observing the decline and restoration of MgH_2 modes at 950 and 1275 cm^{-1} [116]. In the same work, Raman monitoring of hydrogen release from the $LiNH_2$ + 2LiH system provided details on the reaction pathway. The original peaks of $LiNH_2$ at 3262 and 3323 cm^{-1} disappeared at 250 and 400°C and reappeared after annealing the sample in 20 bar H_2 at 340°C. Similar effects were observed for the $NaAlH_4$ + $TiCl_3$ system. The authors noted that laser heating effects might affect the observable temperature of thermal events in the spectra if compared

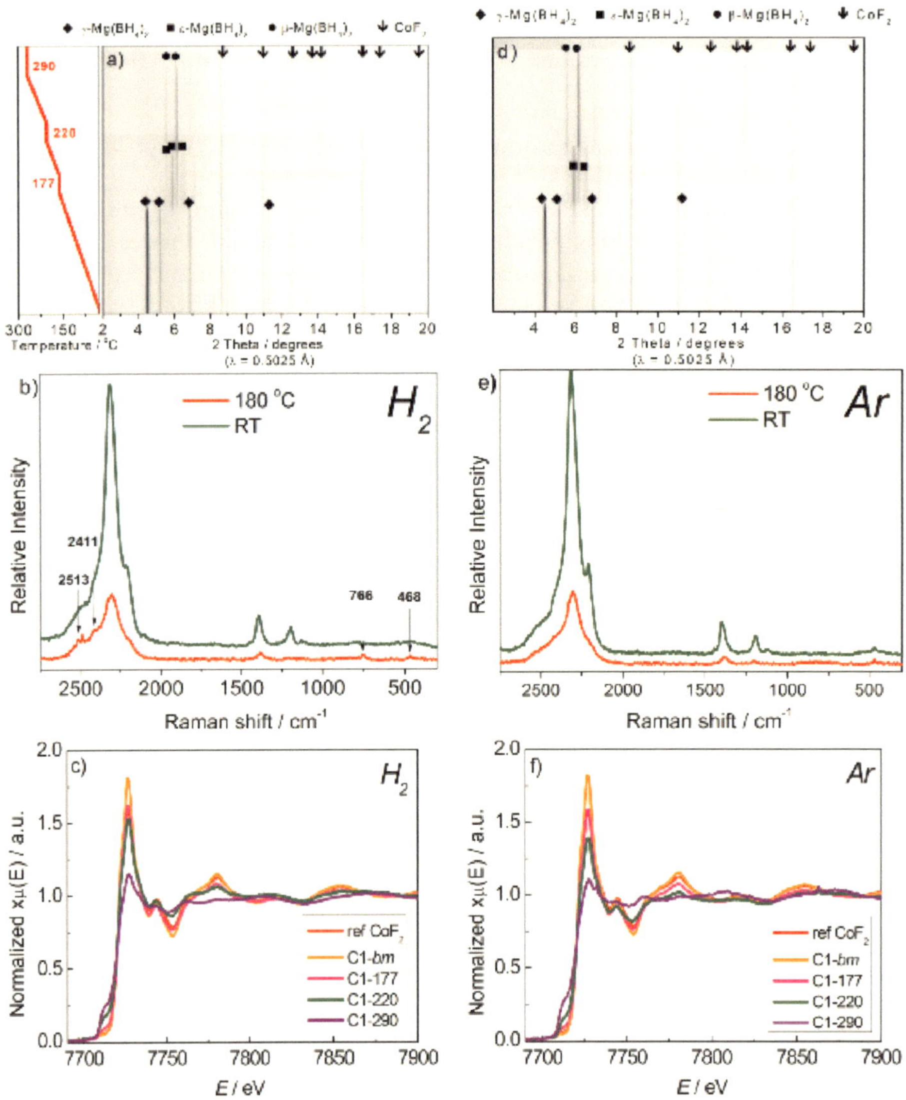

Figure 5.
In situ characterization using (from top to bottom) SR-PXD, Raman, and X-ray absorption (XAS) spectroscopies of the thermal decomposition of Mg(BH4)2 + 2 mol% CoF2 in the RT-300°C range in two different annealing environments: (a–c) 2.5 bar H2 and (d–f) 1 bar Ar [126].

to differential scanning calorimetry (DSC) studies, and that the precise onset temperatures are difficult to determine with Raman. Hino et al. [122] have shown the evolution of quantifiable traces of NH_3 byproduct in the H_2-desorption Raman experiment from the LiH-$LiNH_2$ composite. They have also used the intensity of Raman scattering from NH_3 gas to determine the standard enthalpy and entropy of the decomposition reaction of $LiNH_2$ [123].

Decomposition of CsB_3H_8 was followed by *in situ* Raman at 190°C. In combination with SR-PXRD and FTIR analysis, it was found that $CsBH_4$ is one of the decomposition products of CsB_3H_8 [118]. This result is particularly interesting in view of the fact that $[B_3H_8]^-$, on the contrary, was proposed as the reversible intermediate decomposition product of $[BH_4]^-$ in $Mg(BH_4)_2$ [124].

A study on dehydrogenation and ball-milling of $LiNH_2$-$LiBH_4$-MgH_2 composite found a mixture of $Li_4BN_3H_{10}$ and $Mg(NH_2)_2$ after milling with no evidence of residual $LiNH_2$ or $LiBH_4$. Dehydrogenation onset was observed in the 353–415 K interval, with $Mg(NH_2)_2$ decomposing first, whereas $Li_4BN_3H_{10}$ melted and reacted with MgH_2 to form $Li_2Mg(NH)_2$ and hydrogen gas. Upon cycling at 423 K and 8 MPa H_2 the Raman modes of $Li_4BN_3H_{10}$ reappered, indicating partial reversibility of the composite [125].

Decomposition of $Mg(BH_4)_2$ with Co-based additives was studied by means of a combination of synchrotron X-ray diffraction (XRD) and absorption (XAS) and Raman spectroscopy [126] (**Figure 5**). XRD data showed a decrease in the crystalline phase transition temperature caused by the additives, Raman data showed the formation of an amorphous decomposition phase with vibrations at 2513, 2411, and 766 cm^{-1} in the sample with CoF_2, undetectable in the diffraction patterns, and XAS measurements at the Co K-edge revealed chemical transformation of the additives.

Combined *operando* Raman mass spectrometry methodology was used to study hydrogen evolution from nanoconfined ammonia borane (AB). Among several scaffolds, SPA-15 was found to cause the most prominent effect on the decomposition temperature of AB, whereas Raman spectra indicated a modification in the mechanism of decomposition, suggesting a chemical interaction between hydride and host [127].

4. Conclusion and outlook

The study of materials for hydrogen storage with Raman spectroscopy is a powerful and versatile approach, offering insights into molecular interactions, surface characterization, bonding environments, crystalline structure, phase transitions, formation of solid-state solutions, and reaction mechanisms across a wide spectrum of hydrogen-rich compounds. From probing the subtle shifts in vibrational frequencies of molecular hydrogen and its isotopes to characterizing hydrides and their composites, Raman spectroscopy enables a deeper understanding of hydrogen behavior in various environments. The technique's compatibility with *in situ* and *operando* setups further enhances its utility, allowing for monitoring real-time ad−/absorption and desorption processes. This is a useful property both in the laboratory and in industrial settings, necessary for evaluating the performance and reversibility of hydrogen storage systems. Integration of RS with complementary methods such as X-ray diffraction and spectroscopy, neutron scattering, FTIR spectroscopy, and mass spectrometry opens new avenues for multi-scale characterization, offering insight into hydrogen storage mechanisms in different scales.

Despite its demonstrated potential, Raman spectroscopy remains underutilized, especially in certain areas of hydrogen storage research, such as the study of metallic hydrides, due to the challenges posed by metallic conductivity; however, demonstration studies do exist. Furthermore, the use of isotope substitution offers a promising route to understand reaction pathways and enhance the interpretability of spectral data and could have been employed in a wider research community.

Although not reviewed in this chapter due to the space restrictions, the integration of machine learning (ML) for spectral analysis and the development of single-molecule Raman techniques in combination with tip-enhanced and resonance Raman scattering have the potential to boost the field, enabling automated, high-throughput screening of hydrogen storage materials and unambiguous phase identification, and fundamental insights into the molecular properties, respectively. For the development of ML data analysis, the availability of the spectra libraries is crucial, and hydride compounds do not constitute, for now, a considerable part of the available spectra libraries, like the existing crystallographic databases. For example, the attempts of the author to find spectra of some hydrides in open databases during the preparation of this chapter were not successful.

As the global energy landscape shifts toward sustainable solutions, the role of Raman spectroscopy in guiding the design and optimization of next-generation hydrogen storage systems will become increasingly important.

Author details

Olena Zavorotynska
Department of Mathematics and Physics, University of Stavanger, Stavanger, Norway

*Address all correspondence to: olena.zavorotynska@uis.no

References

[1] Li Q, Yang Y, Yu X, Li H. A 700 W·h·kg−1 rechargeable pouch type lithium battery. Chinese Physics Letters. 2023;**40**(4):048201

[2] Al Ghafri SZS, Munro S, Cardella U, Funke T, Notardonato W, Trusler JPM, et al. Hydrogen liquefaction: A review of the fundamental physics, engineering practice and future opportunities. Energy and Environmental Science. 2022;**15**(7):2690-2731

[3] Mohtadi R, Orimo S-i. The renaissance of hydrides as energy materials. Nature Reviews Materials. 2016;**2**:16091

[4] Jander JH, Chittem PK, Kerscher M, Rausch MH, Wasserscheid P, Fröba AP. Raman spectroscopy for the determination of hydrogen concentration in liquid organic hydrogen carrier systems. International Journal of Hydrogen Energy. 2024;**73**:681-694

[5] George L, Saxena SK. Structural stability of metal hydrides, alanates and borohydrides of alkali and alkali- earth elements: A review. International Journal of Hydrogen Energy. 2010;**35**(11): 5454-5470

[6] Moradi R, Groth KM. Hydrogen storage and delivery: Review of the state of the art technologies and risk and reliability analysis. International Journal of Hydrogen Energy. 2019;**44**(23): 12254-12269

[7] Yang F, Wang T, Deng X, Dang J, Huang Z, Hu S, et al. Review on hydrogen safety issues: Incident statistics, hydrogen diffusion, and detonation process. International Journal of Hydrogen Energy. 2021;**46**(61): 31467-31488

[8] Minuto FD, Rozzi E, Borchiellini R, Lanzini A. Modeling hydrogen storage at room temperature: Adsorbent materials for boosting pressure reduction in compressed H2 tanks. Journal of Energy Storage. 2024;**90**:111758

[9] Zhang L, Allendorf MD, Balderas-Xicohténcatl R, Broom DP, Fanourgakis GS, Froudakis GE, et al. Fundamentals of hydrogen storage in nanoporous materials. Progress in Energy. 2022;**4**(4):042013

[10] Chavan SM, Zavorotynska O, Lamberti C, Bordiga S. H2 interaction with divalent cations in isostructural MOFs: A key study for variable temperature infrared spectroscopy. Dalton Transactions. 2013;**42**(35): 12586-12595

[11] Bordiga S, Lamberti C, Bonino F, Travert A, Thibault-Starzyk F. Probing zeolites by vibrational spectroscopies. Chemical Society Reviews. 2015;**44**(20): 7262-7341

[12] Zhou HL, Yu Y, Zhang HF, Gao T. Structural, vibrational and thermodynamic properties of Mg2 FeH6 complex hydride. The European Physical Journal B. 2011;**79**(3):283-288

[13] Brunel JM. Polysilanes: The grail for a highly-neglected hydrogen storage source. International Journal of Hydrogen Energy. 2017;**42**(36): 23004-23009

[14] Sato T, Orimo S-i. Hydrogen vibration in hydrogen storage materials investigated by inelastic neutron scattering. Topics in Catalysis. 2021; **64**(9):614-621

[15] Lohstroh W, Heere M. Structure and dynamics of borohydrides studied by neutron scattering techniques: A review.

Journal of the Physical Society of Japan. 2020;**89**(5):12

[16] Häussinger P, Lohmüller R, Watson AM. Hydrogen, 1. Properties and Occurence. In: Ullmann's Encyclopedia of Industrial Chemistry. ed. Wiley; 2011. DOI: 10.1002/14356007.a13_297.pub2

[17] Silvera IF. The solid molecular hydrogens in the condensed phase: Fundamentals and static properties. Reviews of Modern Physics. 1980;**52**(2): 393-452

[18] Salumbides EJ, Dickenson GD, Ivanov TI, Ubachs W. QED effects in molecules: Test on rotational quantum states of $\mathbf{H_2}$. Physical Review Letters. 2011;**107**(4):043005

[19] Lamperti M, Rutkowski L, Ronchetti D, Gatti D, Gotti R, Cerullo G, et al. Stimulated Raman scattering metrology of molecular hydrogen. Communications Physics. 2023;**6**(1):67

[20] Nakamoto K. Infrared and Raman Spectra of Inorganic and Coordination Compounds. Hoboken, New Jersey: John Wiley & Sons, Inc.; 2009

[21] Herzberg G, Spinks JWT. Molecular spectra and molecular structure. In: Vol. 1: Spectra of Diatomic Molecules. 2nd ed. Springer, New York, NY: D. VAN Mostrand Company, Inc.; 1963

[22] Spoto G, Vitillo JG, Cocina D, Damin A, Bonino F, Zecchina A. FTIR spectroscopy and thermodynamics of hydrogen adsorbed in a cross-linked polymer. Physical Chemistry Chemical Physics. 2007;**9**(36):4992-4999

[23] Stoicheff BP. High resolution Raman spectroscopy of gases: IX. SPECTRA OF H2, HD, AND D2. Canadian Journal of Physics. 1957;**35**(6):730-741

[24] Bhatnagar SS, Allin EJ, Welsh HL. The Raman spectra of liquid and solid H2, D2, and HD at high resolution. Canadian Journal of Physics. 1962;**40**(1):9-23

[25] Luther F, Sellner GM, Richter M, Stanwix PL. Inherent single-point calibration for quantitative Raman analysis of ortho-Para hydrogen composition. International Journal of Hydrogen Energy. 2025;**102**:778-789

[26] C-s Z, Cohen RE, Mao H-k, Hemley RJ. Raman measurements of phase transitions in dense solid hydrogen and deuterium to 325 GPa. National Academy of Sciences of the United States of America. 2014;**111**(13): 4792-4797

[27] Howie RT, Dalladay-Simpson P, Gregoryanz E. Raman spectroscopy of hot hydrogen above 200 GPa. Nature Materials. 2015;**14**(5):495-499

[28] Li L, Zhang X, Luan Z, Du Z, Xi S, Wang B, et al. Raman vibrational spectral characteristics and quantitative analysis of H2 up to 400°C and 40 MPa. Journal of Raman Spectroscopy. 2018; **49**(10):1722-1731

[29] Zhao D, Yuan D, Zhou H-C. The current status of hydrogen storage in metal–organic frameworks. Energy and Environmental Science. 2008;**1**(2): 222-235

[30] Hadjiivanov KI, Panayotov DA, Mihaylov MY, Ivanova EZ, Chakarova KK, Andonova SM, et al. Power of infrared and Raman spectroscopies to characterize metal-organic frameworks and investigate their interaction with guest molecules. Chemical Reviews. 2021;**121**(3): 1286-1424

[31] Spoto G, Bordiga S, Zecchina A, Cocina D, Gribov EN, Regli L, et al. New

frontier in transmission IR spectroscopy of molecules adsorbed on high surface area solids: Experiments below liquid nitrogen temperature. Catalysis Today. 2006;**113**(1):65-80

[32] Zavorotynska O, Vitillo JG, Spoto G, Zecchina A. Role of extraframework metal sites for hydrogen adsorption into the pores of a zeolite: FT-IR study. International Journal of Hydrogen Energy. 2011;**36**(13):7944-7950

[33] Bénard P, Chahine R. Storage of hydrogen by physisorption on carbon and nanostructured materials. Scripta Materialia. 2007;**56**(10):803-808

[34] Garrone E, Bonelli B, Otero AC. Enthalpy–entropy correlation for hydrogen adsorption on zeolites. Chemical Physics Letters. 2008;**456**(1): 68-70

[35] Murray LJ, Dincă M, Long JR. Hydrogen storage in metal–organic frameworks. Chemical Society Reviews. 2009;**38**(5):1294-1314

[36] Morris L, Trudeau ML, Reed D, Book D, Antonelli DM. High-pressure Raman and calorimetry studies of vanadium(III) alkyl hydrides for Kubas-type hydrogen storage. ChemPhysChem. 2016;**17**(6):822-828

[37] Andalibi MR, Qajar A, Foley HC. Evidence of Nanoconfinement effects in the adsorption of hydrogen on coinage metal complexes dispersed within porous carbon. The Journal of Physical Chemistry C. 2015;**119**(37):21314-21322

[38] Kowalczyk P, Gauden PA, Terzyk AP, Bhatia SK. Thermodynamics of hydrogen adsorption in slit-like carbon Nanopores at 77 K. Classical versus path-integral Monte Carlo simulations. Langmuir. 2007;**23**(7): 3666-3672

[39] Rzepka M, Lamp P, de la Casa-Lillo MA. Physisorption of hydrogen on microporous carbon and carbon nanotubes. The Journal of Physical Chemistry B. 1998;**102**(52):10894-10898

[40] Sumida K, Stück D, Mino L, Chai JD, Bloch ED, Zavorotynska O, et al. Impact of metal and anion substitutions on the hydrogen storage properties of M-BTT metal-organic frameworks. Journal of the American Chemical Society. 2013; **135**(3):1083-1091

[41] Otero Areán C, Nachtigallová D, Nachtigall P, Garrone E, Rodríguez DM. Thermodynamics of reversible gas adsorption on alkali-metal exchanged zeolites—The interplay of infrared spectroscopy and theoretical calculations. Physical Chemistry Chemical Physics. 2007;**9**(12):1421-1437

[42] Kubas GJ. Fundamentals of H2 binding and reactivity on transition metals underlying hydrogenase function and H2 production and storage. Chemical Reviews. 2007;**107**(10): 4152-4205

[43] Hoang TKA, Hamaed A, Moula G, Aroca R, Trudeau M, Antonelli DM. Kubas-type hydrogen storage in V(III) polymers using tri- and Tetradentate bridging ligands. Journal of the American Chemical Society. 2011; **133**(13):4955-4964

[44] Morris L, Trudeau ML, Reed D, Book D, Antonelli DM. Thermodynamically neutral Kubas-type hydrogen storage using amorphous Cr (iii) alkyl hydride gels. Physical Chemistry Chemical Physics. 2015; **17**(14):9480-9487

[45] Rojas-Garcia E, Castañeda-Ramírez AA, Angeles-Beltrán D, López-Medina R, Maubert-Franco AM. Enhancing in the hydrogen storage by

SWCNT/HKUST-1 composites: Effect of SWCNT amount. Catalysis Today. 2022; **394-396**:357-364

[46] Ou JZ, Campbell JL, Yao D, Wlodarski W, Kalantar-zadeh K. In situ Raman spectroscopy of H2 gas interaction with layered MoO3. The Journal of Physical Chemistry C. 2011; **115**(21):10757-10763

[47] Goncharov AF, Struzhkin VV. Raman spectroscopy of metals, high-temperature superconductors and related materials under high pressure. Journal of Raman Spectroscopy. 2003;**34** (7–8):532-548

[48] Schneemann A, White JL, Kang S, Jeong S, Wan LWF, Cho ES, et al. Nanostructured metal hydrides for hydrogen storage. Chemical Reviews. 2018;**118**(22):10775-10839

[49] Adams M, Buckley CE, Busch M, Bunzel R, Felderhoff M, Heo TW, et al. Hydride-based thermal energy storage. Progress in Energy. 2022;**4**(3):27

[50] Saldan I, Schulze M, Pistidda C, Gosalawit-Utke R, Zavorotynska O, Rude LH, et al. Hydrogen sorption in the LiH-LiF-MgB2 system. Journal of Physical Chemistry C. 2013;**117**(33): 17360-17366

[51] Saldan I, Campesi R, Zavorotynska O, Spoto G, Baricco M, Arendarska A, et al. Enhanced hydrogen uptake/release in 2LiH-MgB2 composite with titanium additives. International Journal of Hydrogen Energy. 2012;**37**(2): 1604-1612

[52] Alcantara KS, Boesenberg U, Zavorotynska O, von Colbe JB, Taube K, Baricco M, et al. Sorption and desorption properties of a CaH2/MgB2/CaF2 reactive hydride composite as potential hydrogen storage material. Journal of Solid State Chemistry. 2011;**184**(11): 3104-3109

[53] Grube E, Jensen SRH, Nielsen UG, Jensen TR. Reactivity of magnesium borohydride - metal hydride composites, γ-Mg(BH4)2 -MHx, M = Li, Na, Mg, Ca. Journal of Alloys and Compounds. 2019; **770**:1155-1163

[54] Hayes J, Goudy A. Thermodynamics, kinetics and modeling studies of KH-RbH- and CsH-doped 2LiNH2/MgH2 hydrogen storage systems. International Journal of Hydrogen Energy. 2015; **40**(36):12336-12342

[55] Aoki M, Ohba N, Noritake T, Towata S. Reversible hydriding and dehydriding properties of CaSi: Potential of metal silicides for hydrogen storage. Applied Physics Letters. 2004;**85**(3): 387-388

[56] Beyer DC, Spektor K, Vekilova OY, Grins J, Barros Brant Carvalho PH, Leinbach LJ, et al. Synthesis of BaSiH6 Hydridosilicate at high pressures—a bridge to BaSiH8 Polyhydride. ACS Omega. 2025;**10**(15):15029-15035

[57] Chotard JN, Tang WS, Raybaud P, Janot R. Potassium silanide (KSiH3): A reversible hydrogen storage material. Chemistry-A European Journal. 2011; **17**(44):12302-12309

[58] Zavorotynska O, Saldan I, Hino S, Humphries TD, Deledda S, Hauback BC. Hydrogen cycling in γ-Mg(BH4)2 with cobalt-based additives. Journal of Materials Chemistry A. 2015;**3**(12): 6592-6602

[59] Zavorotynska O, Deledda S, Hauback BC. Kinetics studies of the reversible partial decomposition reaction in Mg(BH4)2. International Journal of Hydrogen Energy. 2016;**41**(23): 9885-9892

[60] Humphries TD, Sheppard DA, Buckley CE. Recent advances in the 18-electron complex transition metal hydrides of Ni, Fe, Co and Ru. Coordination Chemistry Reviews. 2017; **342**:19-33

[61] Parker SF. Spectroscopy and bonding in ternary metal hydride complexes-potential hydrogen storage media. Coordination Chemistry Reviews. 2010;**254**(3–4):215-234

[62] Reilly JJ Jr, Wiswall RH Jr. Reaction of hydrogen with alloys of magnesium and nickel and the formation of Mg2NiH4. Inorganic Chemistry. 1968; 7(11):2254-2256

[63] Unemoto A, Matsuo M, Orimo S-i. Complex hydrides for electrochemical energy storage. Advanced Functional Materials. 2014;**24**(16):2267-2279

[64] Bogdanović B, Schwickardi M. Ti-doped alkali metal aluminium hydrides as potential novel reversible hydrogen storage materials1Invited paper presented at the international symposium on metal–hydrogen systems, les Diablerets, august 25–30, 1996, Switzerland. Journal of Alloys and Compounds. 1997;**253-254**:1-9

[65] Heere M, Zavorotynska O, Deledda S, Sorby MH, Book D, Steriotis T, et al. Effect of additives, ball milling and isotopic exchange in porous magnesium borohydride. RSC Advances. 2018;**8**(49):27645-27653

[66] Saldan I, Frommen C, Llamas-Jansa I, Kalantzopoulos GN, Hino S, Arstad B, et al. Hydrogen storage properties of γ-Mg(BH4)2 modified by MoO3 and TiO2. International Journal of Hydrogen Energy. 2015;**40**(36):12286-12293

[67] Zavorotynska O, Corno M, Pinatel E, Rude LH, Ugliengo P, Jensen TR, et al. Theoretical and experimental study of LiBH4-LiCl solid solution. Crystals. 2012;**2**(1):144-158

[68] Hino S, Fonnelop JE, Corno M, Zavorotynska O, Damin A, Richter B, et al. Halide substitution in magnesium borohydride. Journal of Physical Chemistry C. 2012;**116**(23):12482-12488

[69] Rude LH, Filso U, D'Anna V, Spyratou A, Richter B, Hino S, et al. Hydrogen-fluorine exchange in NaBH4-NaBF4. Physical Chemistry Chemical Physics. 2013;**15**(41):18185-18194

[70] Rude LH, Zavorotynska O, Arnbjerg LM, Ravnsbæk DB, Malmkjær RA, Grove H, et al. Bromide substitution in lithium borohydride, LiBH4-LiBr. International Journal of Hydrogen Energy. 2011;**36**(24): 15664-15672

[71] Zavorotynska O, Corno M, Damin A, Spoto G, Ugliengo P, Baricco M. Vibrational properties of MBH4 and MBF4 crystals (M = Li, Na, K): A combined DFT, infrared, and Raman study. Journal of Physical Chemistry C. 2011;**115**(38):18890-18900

[72] Zavorotynska O, El-Kharbachi A, Deledda S, Hauback BC. Recent progress in magnesium borohydride Mg(BH4)2: Fundamentals and applications for energy storage. International Journal of Hydrogen Energy. 2016;**41**(32): 14387-14403

[73] Herzberg G. Molecular Spectra and Molecular Structure. Vol.2: Infrared and Raman Spectra of Polyatomic Molecules. Princeton, NJ: Van Nostrand; 1945

[74] Rönnebro E, Majzoub EH. Crystal structure, Raman spectroscopy, and ab initio calculations of a new bialkali alanate K2LiAlH6. Journal of Physical Chemistry B. 2006;**110**(51):25686-25691

[75] Leites LA. Vibrational spectroscopy of carboranes and parent boranes and its capabilities in carborane chemistry. Chemical Reviews. 1992;**92**(2): 279-323

[76] Crawford B. Vibrational Intensities. II. The Use of Isotopes. The Journal of Chemical Physics. 1952;**20**(6):977-981

[77] Zavorotynska O, Deledda S, Li GQ, Matsuo M, Orimo S, Hauback BC. Isotopic exchange in porous and dense magnesium borohydride. Angewandte Chemie-International Edition. 2015; **54**(36):10592-10595

[78] Gremaud R, Łodziana Z, Hug P, Willenberg B, Racu AM, Schoenes J, et al. Evidence for hydrogen transport in deuterated $LiBH_4$ from low-temperature Raman-scattering measurements and first-principles calculations. Physical Review B. 2009;**80**(10):100301

[79] Hagemann H, D'Anna V, Rapin J-P, Yvon K. Deuterium−Hydrogen Exchange in Solid Mg(BH4)2. The Journal of Physical Chemistry C. 2010; **114**(21):10045-10047

[80] Sharma M, Sethio D, D'Anna V, Fallas JC, Schouwink P, Cerny R, et al. Isotope exchange reactions in Ca(BH4)2. Journal of Physical Chemistry C. 2015; **119**(1):29-32

[81] Ramzan M, Ahuja R. Ab initio molecular dynamics study of the hydrogen-deuterium exchange in bulk lithiumborohydride (LiBH4). Applied Physics Letters. 2009;**94**(14):2

[82] Boucharat N, Wang D, Bardají EG, Fichtner M, Lohstroh W. Effect of a Ti-based additive on the desorption in isotope-labeled LiB(H,D)4-Mg(H,D)2 nanocomposites. Journal of Physical Chemistry C. 2012;**116**(22): 11877-11885

[83] Lohstroh W, Fichtner M. Rate limiting steps of the phase transformations in Ti-doped NaAlH4 investigated by isotope exchange. Physical Review B. 2007;**75**(18):6

[84] Auffermann G, Müller P, Bronger W. Neutron scattering as a powerful tool in solid state chemistry. Examples from the investigations in crystal structures and properties of metal hydrogen and metal chalcogen compounds. Zeitschrift fur Anorganische und Allgemeine Chemie. 2004;**630**(13–14):2113-2124

[85] Sato T, Sorby MH, Ikeda K, Sato S, Hauback BC, Orimo S. Syntheses, crystal structures, and thermal analyses of solvent-free Ca(AlD4)2 and CaAlD5. Journal of Alloys and Compounds. 2009; **487**(1–2):472-478

[86] Brown WG, Kaplan L, Wilzbach KE. The exchange of hydrogen gas with lithium and sodium borohydrides. Journal of the American Chemical Society. 1952;**74**(5):1343-1344

[87] Mesmer RE, Jolly WL. The exchange of deuterium with solid potassium Hydroborate. Journal of the American Chemical Society. 1962;**84**(11): 2039-2042

[88] Filinchuk Y, Richter B, Jensen TR, Dmitriev V, Chernyshov D, Hagemann H. Porous and dense magnesium borohydride frameworks: Synthesis, stability, and reversible absorption of guest species. Angewandte Chemie, International Edition. 2011; **50**(47):11162-11166

[89] Borgschulte A, Jones MO, Callini E, Probst B, Kato S, Züttel A, et al. Surface and bulk reactions in borohydrides and amides. Energy and Environmental Science. 2012;**5**(5):6823-6832

[90] Gremaud R, Borgschulte A, Friedrichs O, Züttel A. Synthesis mechanism of alkali borohydrides by Heterolytic Diborane splitting. Journal of Physical Chemistry C. 2011;**115**(5): 2489-2496

[91] Borgschulte A, Jain A, Ramirez-Cuesta AJ, Martelli P, Remhof A, Friedrichs O, et al. Mobility and dynamics in the complex hydrides LiAlH4 and LiBH4. Faraday Discussions. 2011;**151**:213-230

[92] Corno M, Pinatel E, Ugliengo P, Baricco M. A computational study on the effect of fluorine substitution in LiBH4. Journal of Alloys and Compounds. 2011; **509**:S679-SS83

[93] Eigen N, Bösenberg U, von Colbe JB, Jensen TR, Cerenius Y, Dornheim M, et al. Reversible hydrogen storage in NaF-Al composites. Journal of Alloys and Compounds. 2009;**477**(1–2):76-80

[94] Humphries TD, Yang J, Mole RA, Paskevicius M, Bird JE, Rowles MR, et al. Fluorine substitution in magnesium hydride as a tool for thermodynamic control. Journal of Physical Chemistry C. 2020;**124**(17):9109-9117

[95] Niu XL, Deng YF, Li X. First-principles study of NaMgH3 by fluorine anion doping for hydrogen storage. Acta Physica Sinica. 2009;**58**(10):7317-7321

[96] Plerdsranoy P, Dansirima P, Jensen TR, Utke R. Hydrogen sorption properties of Li-N-F-H pellets in laboratory and small tank scales. International Journal of Hydrogen Energy. 2023;**48**(73):28435-28444

[97] Yang YJ, Liu YF, Li Y, Gao MX, Pan HG. Fluorine-substituted Mg(BH4) 2•2NH3 with improved dehydrogenation properties for hydrogen storage. Journal of Materials Chemistry A. 2015;**3**(2):570-578

[98] Yin LC, Wang P, Fang ZZ, Cheng HM. Thermodynamically tuning LiBH4 by fluorine anion doping for hydrogen storage:: A density functional study. Chemical Physics Letters. 2008; **450**(4–6):318-321

[99] Yin LC, Wang P, Kang XD, Sun CH, Cheng HM. Functional anion concept: Effect of fluorine anion on hydrogen storage of sodium alanate. Physical Chemistry Chemical Physics. 2007; **9**(12):1499-1502

[100] Renaudin G, Gomes S, Hagemann H, Keller L, Yvon K. Structural and spectroscopic studies on the alkali borohydrides MBH4 (M = Na, K, Rb, Cs). Journal of Alloys and Compounds. 2004;**375**:98-106

[101] Schouwink P, Ley MB, Tissot A, Hagemann H, Jensen TR, Smrčok Ľ, et al. Structure and properties of complex hydride perovskite materials. Nature Communications. 2014;**5**:1-10

[102] Anderson A, Lüty F. Raman scattering, defect luminescence, and phonon spectra of ^{7}Li H, ^{6}Li H, and ^{7}Li D crystals. Physical Review B. 1983;**28**(6): 3415-3421

[103] Dyck W. Lattice dynamics of alkali hydrides and euterides with the NaCl type structure. Journal of Physics C: Solid State Physics. 1981;**14**(29):4193-4215

[104] Barrera GD, Colognesi D, Mitchell PCH, Ramirez-Cuesta AJ. LDA or GGA? A combined experimental inelastic neutron scattering and ab initio lattice dynamics study of alkali metal hydrides. Chemical Physics. 2005;**317**(2–3):119-129

[105] Santisteban JR, Cuello GJ, Dawidowski J, Fainstein A, Peretti HA, Ivanov A, et al. Vibrational spectrum of

magnesium hydride. Physical Review B. 2000;**62**(1):37-40

[106] Tse JS, Klug DD, Desgreniers S, Smith JS, Flacau R, Liu Z, et al. Structural phase transition in CaH_2 at high pressures. Physical Review B. 2007; **75**(13):134108

[107] Wu H, Zhou W, Udovic TJ, Rush JJ, Yildirim T. Structure and vibrational spectra of calcium hydride and deuteride. Journal of Alloys and Compounds. 2007;**436**(1):51-55

[108] Sakai A, Kabutomori T, Misawa T. *In-situ*micro-Raman scattering investigation of $LaNi_{3.4}Co_{1.2}Mn_{0.1}Al_{0.3}$ metal hydride under hydrogen atmosphere. Materials Transactions, JIM. 1999;**40**(9):847-850

[109] Carsteanu AM, Rode M, Zur D, Borgschulte A, Schröter H, Schoenes J. Short-range ordering in $\beta-YH_2+\delta$ and $\beta-YD_2+\delta$ thin films studied by Raman spectroscopy. Physical Review B. 2004; **69**(13):134102

[110] Kierey H, Rode M, Jacob A, Borgschulte A, Schoenes J. Raman effect and structure of YH3 and YD3 thin epitaxial films -: Art. no. 134109. Physical Review B. 2001;**63**(13):11

[111] Aroyo MI, Perez-Mato JM, Capillas C, Kroumova E, Ivantchev S, Madariaga G, et al. Bilbao crystallographic server: I. Databases and crystallographic computing programs. Zeitschrift für Kristallographie - Crystalline Materials. 2006;**221**(1): 15-27

[112] Kroumova E, Aroyo MI, Perez-Mato JM, Kirov A, Capillas C, Ivantchev S, et al. Bilbao crystallographic server: Useful databases and tools for phase-transition studies. Phase Transitions. 2003;**76**(1–2):155-170

[113] Nakamoto K, McKinney MA. Application of the correlation method to Vibrational spectra of C60 and other fullerenes: Predicting the number of IR- and Raman-active bands. Journal of Chemical Education. 2000; **77**(6):775

[114] Gebert F, Willenberg B, van Setten MJ, Bardají EG, Roehm E, Fichtner M, et al. Polarization-dependent Raman spectroscopy of LiBH4 single crystals and Mg(BH4)2 powders. Journal of Raman Spectroscopy. 2011;**42**(9):1796-1801

[115] Carbonniere P, Hagemann H. Fermi resonances of borohydrides in a crystalline environment of alkali metals. Journal of Physical Chemistry A. 2006; **110**(32):9927-9933

[116] Domènech-Ferrer R, Ziegs F, Klod S, Lindemann I, Voigtländer R, Dunsch L, et al. In situ Raman cell for high pressure and temperature studies of metal and complex hydrides. Analytical Chemistry. 2011;**83**(8):3199-3204

[117] Giannasi A, Colognesi D, Ulivi L, Zoppi M, Ramirez-Cuesta AJ, Bardaji EG, et al. High resolution Raman and neutron investigation of Mg(BH4)(2) in an extensive temperature range. Journal of Physical Chemistry A. 2010; **114**(8):2788-2793

[118] Zavorotynska O, Sorby MH, Vitillo JG, Deledda S, Frommen C, Hauback BC. Experimental and computational characterization of phase transitions in CsB3H8. Physical Chemistry Chemical Physics. 2021; **23**(33):17836-17847

[119] Li B, Li YW, Yang KF, Cui QL, Ma YM, Zou GT. Raman evidence of a new high-pressure phase in calcium hydride. Journal of Physics-Condensed Matter. 2007;**19**(22):7

[120] Liu A, Song Y. In situ high-pressure study of sodium amide by Raman and infrared spectroscopies. Journal of Physical Chemistry B. 2011;**115**(1):7-13

[121] Kumar RS, Kim E, Tschauner O, Cornelius AL, Sulic MP, Jensen CM. Pressure-induced structural phase transition in $NaAlH_4$. Physical Review B. 2007;**75**(17):174110

[122] Hino S, Ichikawa T, Ogita N, Udagawa M, Fujii H. Quantitative estimation of NH3 partial pressure in H2 desorbed from the Li–N–H system by Raman spectroscopy. Chemical Communications. 2005;**24**:3038-3040

[123] Hino S, Ogita N, Udagawa M, Ichikawa T, Kojima Y. Thermodynamic properties of lithium amide under hydrogen pressure determined by Raman spectroscopy. Journal of Applied Physics. 15 Jan 2009;**105**(2):023527

[124] Chong M, Karkamkar A, Autrey T, Orimo S, Jalisatgi S, Jensen CM. Reversible dehydrogenation of magnesium borohydride to magnesium triborane in the solid state under moderate conditions. Chemical Communications. 2011;**47**(4): 1330-1332

[125] Hattrick-Simpers JR, Maslar JE, Niemann MU, Chiu C, Srinivasan SS, Stefanakos EK, et al. Raman spectroscopic observation of dehydrogenation in ball-milled LiNH2-LiBH4-MgH2 nanoparticles. International Journal of Hydrogen Energy. 2010;**35**(12):6323-6331

[126] Zavorotynska O, Deledda S, Vitillo J, Saldan I, Guzik M, Baricco M, et al. Combined X-ray and Raman studies on the effect of cobalt additives on the decomposition of magnesium borohydride. Energies. 2015;**8**(9):9173

[127] Valero-Pedraza MJ, Gascón V, Carreón MA, Leardini F, Ares JR, Martín A, et al. Operando Raman-mass spectrometry investigation of hydrogen release by thermolysis of ammonia borane confined in mesoporous materials. Microporous and Mesoporous Materials. 2016;**226**:454-465

Chapter 6

Advanced Applications of 3D Raman Imaging in Biomedicine

Xiaoyu Zhao

Abstract

Label-free, super-multiplexed Raman profiling of three-dimensional (3D) biological samples offers transformative potential across biomedical research and clinical diagnostics. Unlike conventional imaging methods, Raman spectral imaging enables simultaneous, preparation-free detection of a broad range of biomolecules, including nucleic acids, proteins, and diverse lipid subtypes in their native spatial context. However, the full utility of Raman spectroscopy has long been constrained by challenges in signal acquisition and spectral unmixing, which limit both molecular specificity and quantitative accuracy. Recent advances in reference-based computational unmixing techniques and spectral normalization now overcome these barriers, enabling not only improved quantification but also precise spatial mapping of biomolecular distributions in complex 3D systems. The accuracy of Raman-derived molecular maps can be validated by correlative immunostaining, which confirms the spatial fidelity of certain biomolecular patterns. A particularly promising application is the *in situ* characterization of the extracellular matrix (ECM), a dynamic microenvironment critical for tissue development, repair, and intercellular signaling. Traditional histological and molecular approaches often fail to preserve spatial integrity while detecting ECM-associated signaling molecules. In contrast, 3D Raman imaging uniquely captures both spatial and temporal heterogeneity of ECM components and secreted factors, offering mechanistic insights previously inaccessible. This advancement represents a major leap in the label-free, high-content analysis of live and fixed biological systems.

Keywords: 3D Raman imaging, ECM characterization, label-free imaging, live imaging, super-multiplexed imaging

1. Introduction

Monitoring the biomolecular composition of human cells and tissues is essential, as key physiological processes, such as tissue maturation, organ development, and disease progression, are often characterized by the emergence, disappearance, or fluctuation of specific biomolecules. Fundamental cellular activities, including differentiation, apoptosis, and proliferation, are closely associated with dynamic changes in molecular content and concentration. To capture these variations and assess changes in cellular density and biomolecular profiles, a wide range of imaging techniques have been developed, each offering unique insights into the structural and functional states of biological systems.

A variety of advanced imaging modalities have been developed to explore the spatial and molecular organization of biological samples in three dimensions. While each method offers unique advantages, they also present specific limitations, particularly in the context of label-free, high-content, and non-destructive analysis of intact biological systems: Cryo-electron tomography (cryo-ET) provides high-resolution 3D imaging of cellular ultrastructure, but lacks the ability to differentiate molecular components and requires cryo-preservation [1]. Mass spectrometry imaging (MSI) offers label-free, high-specificity molecular detection but is limited to thin sections and 2D analysis [2]. Confocal fluorescence microscopy, commonly used for 3D tissue imaging, suffers from spectral overlap and requires labeling, which may alter cellular processes and obscure the natural state [3]. Light sheet fluorescence microscopy (LSFM) enables rapid volumetric imaging but also depends on fluorescent labels and requires optical clearing, limiting its multiplexing capabilities [4]. Expansion microscopy (ExM) provides nanoscale resolution but necessitates extensive sample preparation and labeling, making it incompatible with live-cell imaging [5]. Label-free infrared (IR) imaging offers chemical information but struggles with spatial resolution and water interference, particularly in thicker samples [6].

To characterize biological samples and monitor their compositional and structural changes, 3D confocal Raman spectroscopic imaging serves as a powerful tool. This label-free, inelastic light scattering optical technique enables the visualization of the spatial distribution of biomolecular structures and cellular morphology in both fixed and living tissues. Raman spectroscopy measures the characteristic vibrational modes of chemical bonds within various cellular components (e.g., nucleic acids, proteins, lipids, and cytochrome c), allowing distinct Raman spectra to be obtained for nearly all biomolecules. Compared to fluorescence imaging (~500 cm^{-1}), Raman imaging offers much narrower vibrational peaks (~10 cm^{-1}) and does not suffer from the "color barrier" issue, making it fundamentally advantageous for multiplexed imaging. In addition, Raman imaging benefits from the inherently low absorption cross-section of water, which stands in contrast to techniques such as infrared (IR) spectroscopy, where strong water absorption can obscure important spectral features [7]. This property gives Raman spectroscopy a significant advantage in the analysis of aqueous and hydrated biological samples, including live cells, tissues, and organoids. Moreover, Raman is uniquely compatible with confocal optical configurations, enabling subcellular spatial resolution and three-dimensional sectioning. This allows for the detailed mapping of spatial and biochemical heterogeneity at the single-cell level, making it a powerful tool for characterizing the pathophysiological microenvironment, cellular phenotypes, and molecular gradients in complex biological systems [8].

With a well-calibrated reference spectral library generated using liquid tissue phantoms and a standardized analytical method for spectral unmixing, 3D quantitative biomolecular profiles with subcellular resolution can be derived from Raman signals of human tissues or organoid samples [7]. This enables the simultaneous imaging and analysis of proteins, lipids, and metabolites. As a result, a comprehensive compositional and structural analysis of 3D biological systems can be achieved in a single multiplexed Raman scan, potentially targeting hundreds of biomolecules in samples up to half a millimeter thick, without the need for sample preparation. Importantly, living tissue organoids remain intact after Raman imaging, enabling continuous culture for further analysis [9].

Technique	Label-free	Resolution	Sample preparation	3D/2D	Key advantages	Key limitations
Cryo-Electron Tomography (cryo-ET)	No	Nanometer (~1–2 nm)	Cryo-preservation required	3D	High-resolution structural imaging of cellular ultrastructure and complexes	No chemical specificity; complex sample prep; low throughput
Mass spectrometry imaging (MSI)	Yes	Moderate (~10–50 μm)	Thin sections; destructive	2D	High molecular specificity; wide analyte coverage; label-free	Limited to 2D; requires thin samples; destructive; low resolution
Confocal fluorescence microscopy	No	High (~200 nm)	Fluorescent labeling required	3D	High-resolution optical sectioning; live-cell imaging	Label-dependent; spectral overlap limits multiplexing; perturbative
Light sheet fluorescence microscopy (LSFM)	No	High (~200 nm)	Fluorescent labeling required	3D	Fast volumetric imaging; minimal phototoxicity; suitable for large samples	Label-dependent; optical clearing required; multiplexing limits
Expansion microscopy (ExM)	No	Nanoscale (~20–50 nm)	Sample expansion, labeling required	3D	Achieves super-resolution with conventional microscopes; spatial fidelity	Requires extensive processing; labeling needed; not for live cells
Label-free infrared (IR) imaging	Yes	Moderate (~5–10 μm)	Minimal sample prep, no labeling	2D	Non-destructive; label-free chemical content analysis	Poor resolution; interference from water; challenges with thick samples
Raman spectroscopic imaging	Yes	Moderate (~250-350 nm)	Minimal sample prep, no labeling	3D	Label-free, non-destructive; multiplexed detection; live/fixed samples	Weak signal; fluorescence background; requires advanced unmixing technique [8]

Table 1.
Comparison of advanced imaging techniques for biological sample analysis.

Combined with Raman calibration bio-analytical techniques, Raman imaging overcomes the limitations of other biomolecular characterization methods shown in **Table 1**. In this chapter, we demonstrate the utility of Raman imaging by applying advanced spectral unmixing techniques to human intestinal organoids, human brain organoids, human breast cancer organoids, and real cartilage samples. Additionally, we highlight the application of Raman imaging in unraveling the composition of highly complex drug formulations, such as gel patches, showcasing its use in the field of generic drug reverse engineering.

2. Case studies: Application of Raman profiling in the field of biomedicine

2.1 Raman profiling of human intestinal organoid

Human intestinal organoids serve as a physiologically relevant 3D model for studying gut development, metabolism, barrier function, and disease. Using 3D confocal Raman spectroscopic imaging, we acquired volumetric biochemical maps of human intestinal organoids without any labeling or sample processing, as shown in **Figure 1** and Video 1, http://bit.ly/3J6jy59. Distinct Raman spectral signatures were assigned to Hyaluronan, Collagen, Laminin, Glycogen, Cholesterol, Saturated Lipid, Monounsaturated Lipid, and Unsaturated Lipid. Spectral unmixing, supported by a calibrated reference library, revealed zonal variation in lipid saturation, glycogen storage, ECM distribution, and the functional compartmentalization of the intestinal epithelium and its underlying matrix. These findings align well with known intestinal epithelial cell lineage hierarchies and validate Raman imaging as a powerful method for capturing the compositional architecture of the intestinal mucosa *in vitro*.

2.2 Raman imaging for quality control and drug analysis in human brain organoids

Human organoids represent a transformative model system for drug screening, disease modeling, and personalized therapy. However, ensuring the quality and reproducibility of organoid models remains a critical challenge. Conventional evaluation methods—such as immunostaining or biochemical assays—are typically invasive, endpoint-specific, and destructive, making them unsuitable for real-time, longitudinal monitoring of living samples. In this context, 3D Raman spectroscopic

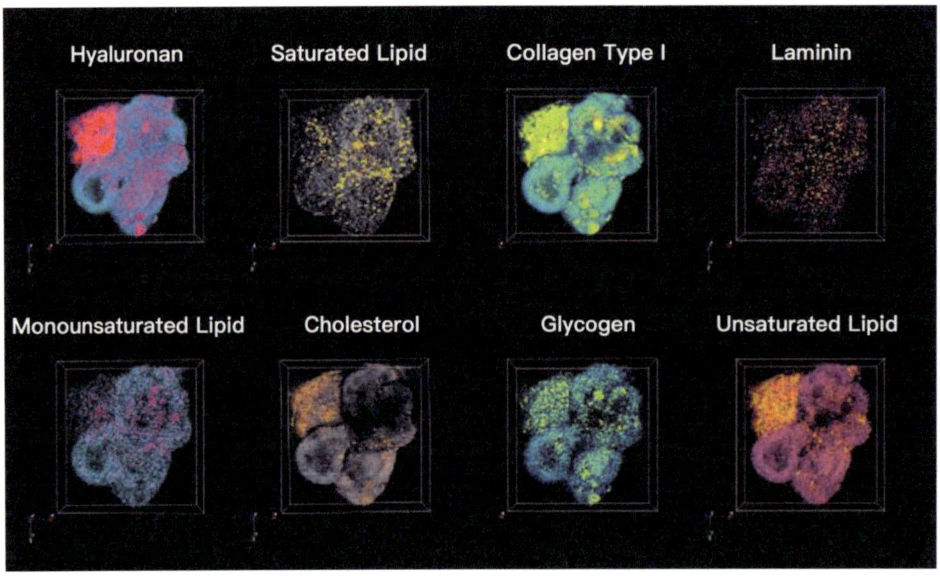

Figure 1.
Raman profiling of fixed human intestinal organoid. Figure reproduced with permission from SuperVision Medicine Co., Ltd.

imaging emerges as a powerful, non-invasive, and label-free technology uniquely suited for organoid quality control.

Due to its ability to detect intrinsic vibrational signatures of biomolecules, Raman imaging allows direct molecular characterization without any staining or labeling, and can be applied to both live and fixed samples. As a case study shown in **Figure 2**, we applied 3D Raman imaging to monitor fixed human brain organoids over a 10-day period. During storage, gradual evaporation of phosphate-buffered saline (PBS) resulted in the formation of PBS salt crystals within the organoid tissue. In **Figure 2**, Raman volumetric imaging at day 2, day 5, day 7, and day 10 enabled clear visualization of the 3D spatial distribution of PBS crystals, which would be difficult or impossible to detect using conventional histological or optical methods. However, since PBS has a well-characterized Raman spectral signature, it could be easily unmixed from the complex organoid background using a reference-based spectral deconvolution approach.

This example illustrates a key strength of Raman imaging: as long as a pure substance is available, its spectral fingerprint can be acquired and used to detect, localize, and quantify it within complex biological matrices. This principle extends to drug molecules and their metabolites, enabling Raman imaging to be used for *in situ* pharmacokinetic analysis [7], drug penetration assessment, and metabolite localization within organoids or tissues without the need for chemical labeling or destructive processing.

Thus, 3D Raman imaging not only solves the unmet need for non-destructive organoid quality control but also offers a unique platform for high-resolution drug distribution studies, batch-to-batch consistency verification, and longitudinal metabolic monitoring in organoid-based drug screening workflows. Its high-content, label-free nature makes it particularly suitable for personalized medicine, where maintaining the integrity and viability of patient-derived organoids is essential [8].

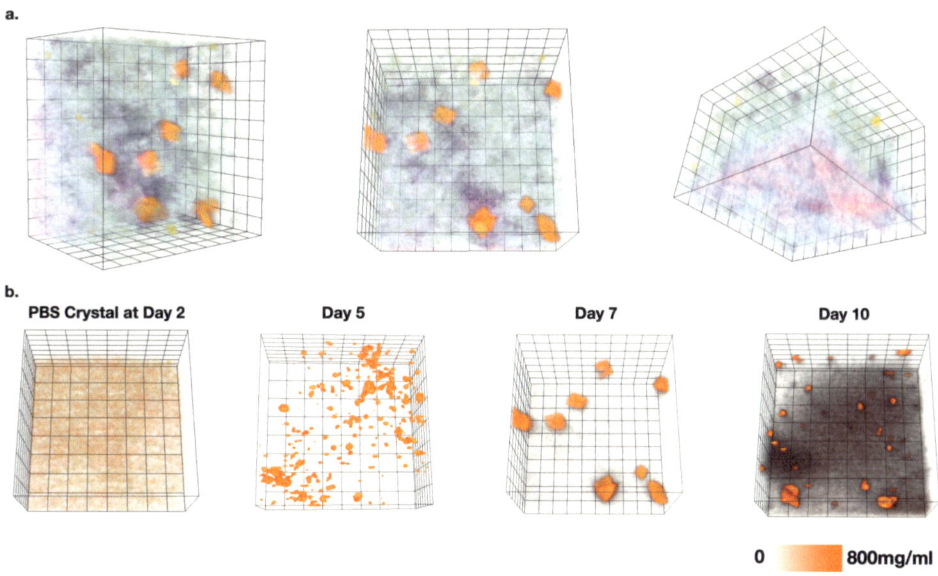

Figure 2.
Raman profiling of human brain organoids. (a) Brain organoid profiling at day 7. (b) PBS crystal formation within brain organoids at day 2, day 5, day 7, and day 10. Figure reproduced with permission from SuperVision Medicine Co., Ltd.

2.3 Raman profiling of human breast cancer orgnoid

Breast cancer organoids recapitulate the cellular and molecular heterogeneity of tumors and serve as a powerful platform for functional profiling of cancer subtypes [8]. Using super-multiplexed 3D Raman spectroscopic imaging, we conducted high-resolution, label-free molecular phenotyping of breast cancer organoids undergoing mesenchymal-to-epithelial transition (MET), as shown in **Figure 3** and Video 2, http://bit.ly/3J6jy59. Distinct Raman spectral signatures were identified for key extracellular matrix (ECM) components, including collagen, fibronectin, elastin, glycogen, and laminin, as well as for various lipid subtypes, such as saturated, monounsaturated, and polyunsaturated lipids. The spatial distribution and relative abundance of these biomolecules provided insights into both epithelial organization and stromal remodeling. These findings demonstrate the capability of Raman imaging to non-invasively distinguish molecular subtypes based on their biochemical composition and suggest its potential for tracking therapeutic responses with subcellular resolution.

2.4 Applications in living human cartilage tissue

Cartilage is a dense, ECM-rich tissue with complex biochemical gradients that are difficult to characterize with conventional methods. Raman spectroscopic imaging was used to quantify and spatially resolve the distribution of collagen types, proteoglycans, elastin, and mineral components in real human articular cartilage. Raman maps revealed zone-specific variation in collagen crosslinking and sulfated

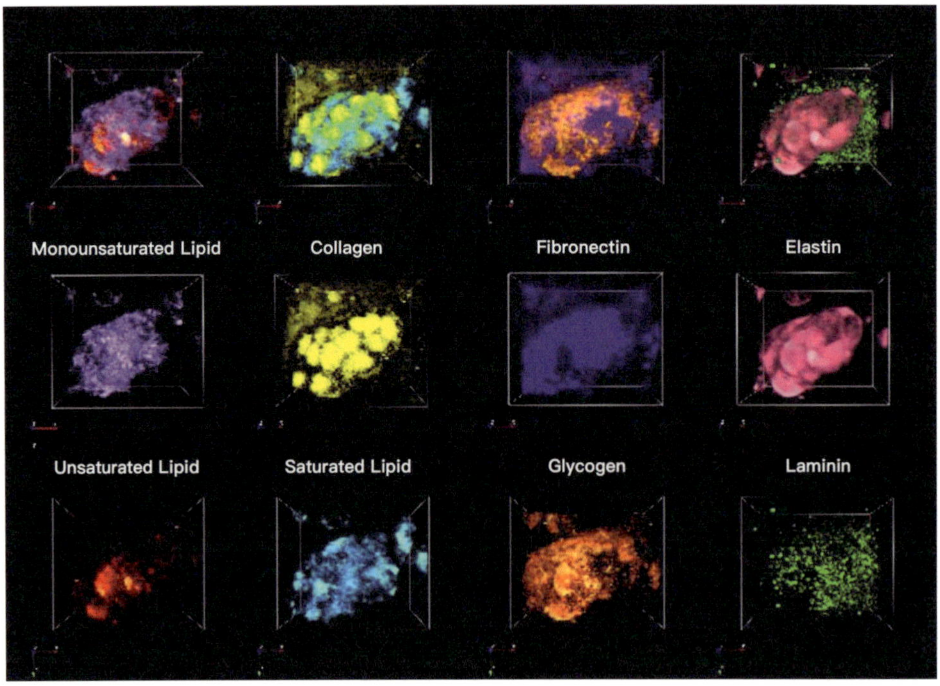

Figure 3.
Raman phenotyping of human breast cancer organoid. Figure reproduced with permission from SuperVision Medicine Co., Ltd.

glycosaminoglycans (sGAGs) from the superficial zone to the deep zone. These signatures correlate with mechanical gradients and structural integrity [9]. Importantly, Raman enabled non-destructive profiling of cartilage degeneration in osteoarthritic samples, identifying early molecular changes (e.g., collagen degradation and lipid infiltration) not detectable by histology alone. This label-free technique thus offers potential in early diagnosis and monitoring of joint disease. Raman imaging can also track the biomolecular changes during cartilage maturation. **Figure 4** shows the changes in the 3D distribution of aggrecan, collagen type, elastin, laminin, glycogen, and saturated lipid from human engineered cartilage tissues *in vitro* at weeks 3, 6, 9, and 15, *in vivo* at week 15, and the real human cartilage.

2.5 Reverse engineering of complex drug formulations

In pharmaceutical development, reverse engineering of complex formulations, such as drug-laden gels, nanocarriers, and transdermal patches, is essential for generic

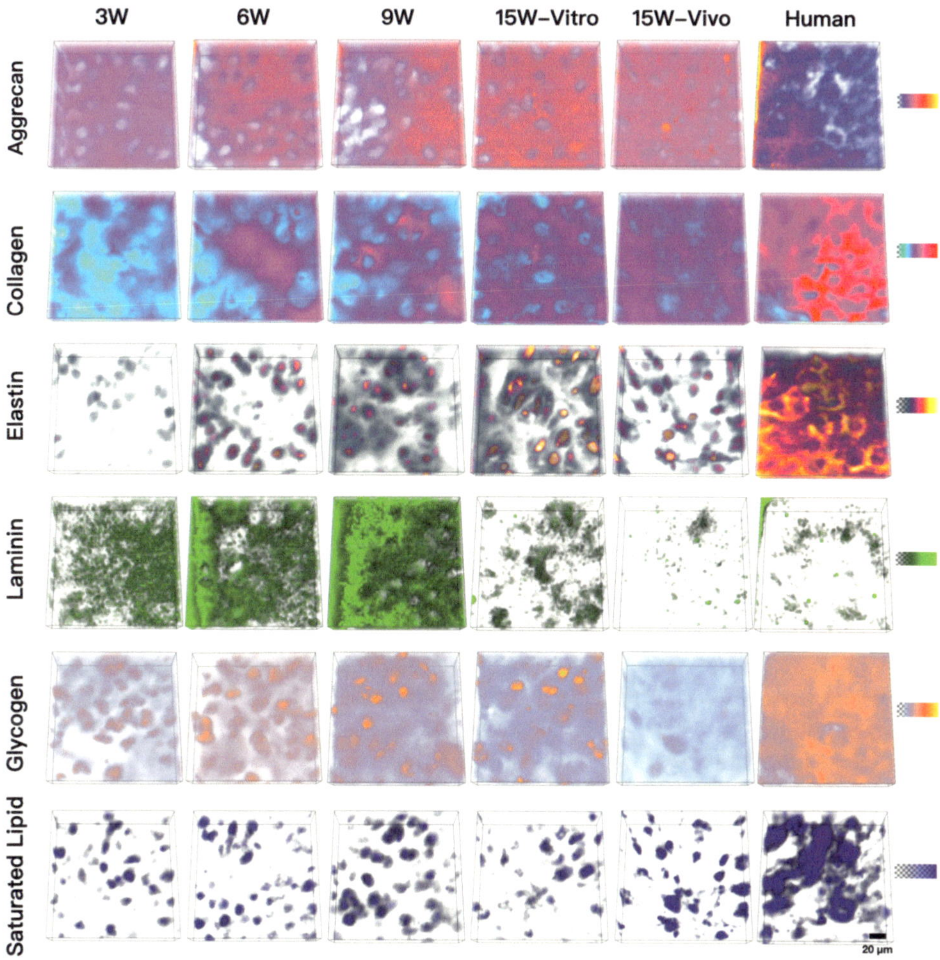

Figure 4.
Raman profiling of cartilage maturation process from week 3 to week 15. Figure reproduced with permission from SuperVision Medicine Co., Ltd.

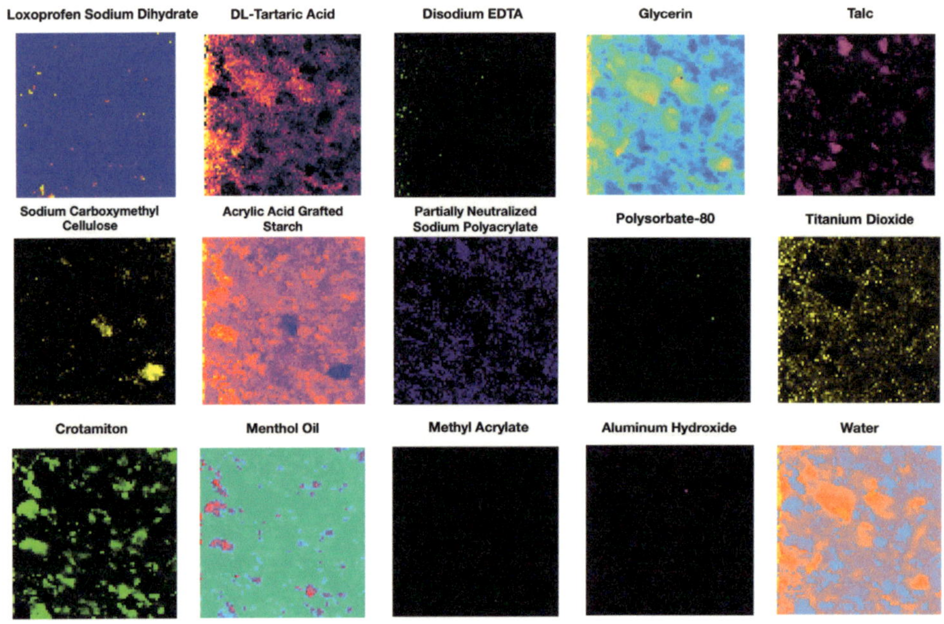

Figure 5.
Raman-based reverse engineering of a loxoprofen sodium gel patch. Figure reproduced with permission from SuperVision Medicine Co., Ltd.

drug development and regulatory approval. These systems often involve multilayered architectures and heterogeneous distribution of active pharmaceutical ingredients (APIs), excipients, and polymer matrices, making spatially resolved, component-level analysis critical.

Raman imaging, with its label-free, non-destructive nature and molecular specificity, offers a powerful tool to meet this need. We applied 2D Raman imaging to analyze highly complex gel patch formulations. By acquiring reference spectra from pure APIs, polymers, surfactants, and lipid carriers, and applying spectral unmixing techniques, we reconstructed the chemical architecture of the patches *in situ*, without the need for extraction or staining.

This enabled precise localization of APIs, detection of excipient crystallization, and identification of inhomogeneities in drug dispersion and polymer blending. Notably, Raman imaging preserved spatial integrity while revealing microstructural features inaccessible to conventional analytical methods. These capabilities make Raman spectroscopy particularly valuable for formulation optimization, stability assessment, and demonstrating bioequivalence in generic drug approval workflows. **Figure 5** presents the Raman-based reverse engineering of a loxoprofen sodium gel patch, demonstrating the capability of Raman spectroscopy to spectrally unmix and identify all major chemical constituents within the formulation. This analysis highlights the potential of Raman spectroscopy for non-destructive, label-free characterization of complex topical pharmaceutical preparations. The identified components and their respective functions in the gel patch are as follows:

1. Loxoprofen Sodium Dihydrate: The active pharmaceutical ingredient (API), a nonsteroidal anti-inflammatory drug (NSAID) used for analgesic and anti-inflammatory effects.

2. DL-Tartaric Acid: A pH adjuster and stabilizer, also used to enhance solubility.

3. Disodium EDTA: A chelating agent that binds divalent metal ions, improving formulation stability and preventing degradation.

4. Glycerin: A humectant that retains moisture and maintains the gel's softness and skin adhesion.

5. Talc: A filler and consistency enhancer; may also aid in moisture control.

6. Sodium Carboxymethyl Cellulose: A thickening agent that provides viscosity and helps form the gel matrix.

7. Acrylic Acid Grafted Starch: A bioadhesive and thickening polymer that supports gel structure and skin adherence.

8. Partially Neutralized Sodium Polyacrylate: A superabsorbent polymer that contributes to gel viscosity and moisture retention.

9. Polysorbate-80: A nonionic surfactant and emulsifier that stabilizes hydrophobic and hydrophilic components.

10. Titanium Dioxide: An opacifier and UV protector; also used for its whitening properties in topical formulations.

11. Crotamiton: An antipruritic agent that can enhance patient comfort.

12. Menthol Oil: Provides a cooling sensation and mild analgesic effect; also enhances skin penetration of the API.

13. Methyl Acrylate: A polymer precursor or monomer used in the matrix, typically present as a polymerized residue.

14. Aluminum Hydroxide: A pH buffering agent and mild astringent; may also contribute to stability.

15. Water: The primary solvent, serving as the carrier for dissolved and dispersed components.

This spectral deconvolution highlights the feasibility of label-free Raman imaging for comprehensive molecular analysis of complex pharmaceutical formulations. Such capability holds significant potential for applications in quality control, counterfeit drug detection, and formulation optimization. Moreover, this approach demonstrates the promise of Raman profiling in the characterization of creams, microspheres, and pellets—formulations that are inherently more complex than conventional solid tablets due to their multicomponent and heterogeneous nature.

3. From semi-quantitative to fully quantitative Raman imaging

For many years, Raman imaging of biological samples has been regarded as a semi-quantitative technique, offering insights into molecular identity and distribution, but lacking a standardized method for absolute quantification. A significant breakthrough occurred in 2023, when the group led by Prof. Molly Stevens developed a robust strategy to transform spontaneous Raman imaging into a fully quantitative modality [7].

This advancement relies on the principle that Raman scattering intensity at each wavenumber is directly proportional to the molecular concentration of specific vibrational bonds. Thus, given a well-calibrated system and known reference spectra, it becomes possible to calculate the absolute abundance of multiple molecular species across a 3D biological volume. The Stevens group introduced a spectral normalization step by referencing each voxel's Raman spectrum to the intensity of the 3300 cm^{-1} peak, associated with water OH stretching. Given that water is ubiquitous and uniformly distributed in hydrated tissues, this internal normalization allows correction for depth-dependent scattering and acquisition noise. The key assumption here is that water content within each voxel remains relatively constant (e.g., ~5 fL), an assumption generally valid in osmotically balanced biological samples [7].

With this strategy, spontaneous Raman imaging can achieve voxel-level absolute quantification of proteins, lipids, nucleic acids, and other molecules in tissues and organoids. This method has been validated in liver organoids and mammary epithelial organoids models, with Raman-derived concentrations correlating with orthogonal biochemical assays and immunostaining [7, 8].

This approach circumvents key limitations of stimulated Raman scattering (SRS), which, although offering higher signal intensity, is limited to narrow spectral windows and cannot resolve complex overlapping signatures of similar biomolecules [10]. In contrast, spontaneous Raman can capture the full vibrational spectrum with high resolution (~1 cm^{-1}), enabling the separation of closely related protein or lipid species, such as Collagen Type I vs. Collagen Type II, or L-Alanine versus DL-Alanine. Enhanced techniques such as Surface-Enhanced Raman Spectroscopy (SERS) and Tip-Enhanced Raman Spectroscopy (TERS) are well-suited for the detection of specific nanoparticulate substances; however, their limited detection area and challenges in obtaining reproducible images render them less applicable to comprehensive biological imaging [11, 12]. These limitations, combined with their narrow spectral acquisition, largely explain why SERS and TERS have not been widely adopted for fully quantitative applications in complex biological systems.

Furthermore, integration of electron-multiplying CCD (EMCCD) detectors and real-time unmixing algorithms allows rapid acquisition of 0.01 to 0.05 seconds per spectrum, which is sufficient for high-throughput quantification in biological workflows [13]. These developments push Raman imaging from qualitative visualization toward a versatile platform for non-invasive molecular quantification in research and clinical applications.

4. Raman imaging for label-free *in situ* cell typing

Cell typing is foundational for understanding tissue architecture, immune responses, disease progression, and therapeutic targeting. Traditional methods such as fluorescence-activated cell sorting (FACS), immunohistochemistry (IHC), and single-cell RNA sequencing rely on labeling and often require tissue dissociation, which results in the loss of spatial context. In contrast, Raman imaging provides a label-free alternative capable of performing *in situ* cell typing directly within intact tissues and organoids, preserving spatial integrity and morphological context [13].

DOI: http://dx.doi.org/10.5772/intechopen.1012300

Unlike flow cytometry, where cells are analyzed in suspension and sorted based on fluorescence intensity, Raman imaging leverages intrinsic biochemical fingerprints of cells for classification [14]. After FACS isolation, label-free Raman reference spectra can be acquired from purified cell types, and these profiles can be used to deconvolve heterogeneous tissues using unmixing algorithms. This approach retains both biochemical and positional information.

4.1 Current challenges in cell typing

Conventional cell typing techniques face several constraints, as shown in **Table 2**:

1. *Label dependency*: Fluorescent or antibody-based labeling is required in flow cytometry and immunofluorescence, which can alter cellular physiology and is time-consuming [15].

2. *Multiplexing limitations*: Fluorescence-based platforms are generally limited to fewer than eight simultaneous targets due to spectral overlap and fluorescence bleed-through [16].

3. *Loss of spatial context*: Methods such as flow cytometry and single-cell RNA sequencing require dissociation of tissues, erasing the tissue architecture and microenvironmental context [3].

4. *Sample destruction*: Techniques like RNA-seq and mass cytometry (CyTOF) require cell lysis or fixation, preventing further analysis of the same sample [17].

5. *Subcellular resolution*: Most conventional methods cannot achieve subcellular localization, which is critical for understanding nuclear–cytoplasmic dynamics or intracellular biomolecular distributions [18].

6. *Marker discovery bottleneck*: Identification of novel or rare cell types is constrained by the need to define specific markers, which may be unknown or poorly characterized [11].

Feature	Flow cytometry	Label-free Raman imaging
Sample format	Cells in suspension	Cells/tissues *in situ*
Labels required	Yes (fluorescent)	No
Multiplexing	Limited (~4–8 markers)	Very high (>10 components)
Spatial context	No	Yes (3D + subcellular)
Cell sorting	Yes (FACS)	No (imaging only)
Destructive	Yes	No
Subcellular resolution	No	Yes
Marker discovery required	Yes	No
Cost and preparation	High	Low

Table 2.
Comparison between flow cytometry and Raman spectroscopy in cell typing.

4.2 Raman imaging as a label-free alternative

Raman spectroscopy provides a fundamentally different approach by detecting intrinsic vibrational signatures of biomolecules, such as lipids, proteins, nucleic acids, and metabolites, without the need for any labels or staining. Each cell type exhibits a unique Raman spectral fingerprint based on its biochemical composition, enabling label-free, non-destructive classification of multiple cell types simultaneously [8].

Raman imaging operates *in situ* and is compatible with three-dimensional biological systems such as tissues and organoids. This preserves tissue architecture and enables spatially resolved cell classification within their native microenvironment. Furthermore, Raman imaging is confocal-compatible, allowing subcellular resolution profiling of biomolecular distributions—a feature rarely achieved by conventional methods [7].

4.3 Complementarity with FACS for reference spectra acquisition

A synergistic workflow can be established by combining FACS and Raman imaging. After isolating pure cell populations *via* FACS, the cells can be imaged by Raman microscopy to generate reference spectral libraries. These reference spectra, devoid of fluorescent labels, can then be used to deconvolve complex tissue samples *via* spectral unmixing algorithms, enabling classification of heterogeneous populations without further labeling [19]. This approach transforms Raman imaging into a quantitative phenotyping tool that links well-characterized reference spectra with complex, intact biological tissues—offering new opportunities for personalized diagnostics, cancer subtype classification [8], and immunophenotyping.

4.4 From 2D to 3D cell typing

Unlike traditional immunofluorescence or histological staining, which are often restricted to 2D thin sections, Raman imaging supports full 3D molecular profiling. This allows for:

i. Visualization of cell–drug and cell–matrix interactions within tissue or organoids [7].

ii. Monitoring of functional states such as activation, senescence, or apoptosis based on spectral features [19].

iii. *In situ* tracking of dynamic transitions such as epithelial–mesenchymal transformation or immune activation [8].

iv. Such capability is especially valuable for studying tumor–immune niches, stem cell environments, and neural circuits, where spatial architecture influences function [8, 19].

4.5 True supermultiplexing and functional profiling

Fluorescence-based imaging is fundamentally constrained by the number of distinguishable dyes, but Raman imaging can distinguish 20 or more biomolecular targets simultaneously due to the narrow linewidth of Raman peaks (~10 cm^{-1})

and the broad spectral range (~400–3500 cm^{-1}) [8]. Cell types are identified not by specific surface markers, but by their composite molecular signature, allowing comprehensive phenotyping even in the absence of known biomarkers [7].

Moreover, Raman spectra reflect not just identity, but functional and metabolic state. For example, shifts in cytochrome c, lipid composition, or nucleic acid content can be used to infer: (i) cellular metabolism, (ii) proliferative status [20], (iii) apoptotic or stress responses, and (iv) Differentiation state [7, 19]. This enables functional cell typing beyond simple phenotypic labeling.

4.6 Operational simplicity and cost-effectiveness

Raman imaging is non-destructive, allowing the same sample to be reused for follow-up assays such as transcriptomics, proteomics, or pathology. Unlike immunolabeling protocols that require multiple washing, blocking, and incubation steps, Raman imaging: (i) requires no specialized reagents [9], (ii) has low running cost, (iii) offers real-time acquisition at the pixel level (typically 0.01–0.3 seconds per spectrum), and (iv) can be integrated into automated workflows for high-throughput analysis [13, 20].

Furthermore, since no chemical or physical alteration is applied to the sample, Raman imaging preserves true biological fidelity—a critical factor in drug screening, diagnostics, and longitudinal studies.

5. Limitations and outlook

Raman spectroscopic imaging offers unparalleled capabilities for label-free, super-multiplexed, and 3D molecular imaging. However, several limitations still hinder its broader adoption, particularly in high-throughput and clinical settings.

First, spontaneous Raman scattering generates inherently weak signals, requiring long acquisition times that limit real-time imaging. Stimulated Raman scattering (SRS) addresses this by enhancing signal intensity up to 10^4-fold, but at the cost of spectral breadth. SRS typically captures only a few Raman shifts per scan, reducing its ability to distinguish structurally similar biomolecules. Its technical complexity—demanding synchronized pulsed lasers, high-speed modulation, and custom optics—also increases cost and limits accessibility. Furthermore, the narrow spectral range of SRS constrains quantitative analysis due to fewer spectral variables for unmixing and curve fitting. In contrast, spontaneous Raman allows full-spectrum acquisition with high spectral resolution, enabling detailed molecular profiling. While slower for imaging, single-point spectra (0.01–0.5 s/pixel) are sufficient for real-time quantification, such as blood glucose monitoring. Hardware advances like electron-multiplying CCDs (EMCCDs) can reduce acquisition time, but their performance and cost vary widely, requiring careful selection for sensitive 3D imaging.

Second, autofluorescence from biological samples can obscure weak Raman signals. Near-infrared (785–1064 nm) excitation mitigates fluorescence but reduces Raman intensity, resolution, and detector sensitivity. Conversely, 532 nm excitation offers an optimal balance and is widely used due to strong signals, good resolution, and minimal phototoxicity. High-performance Raman systems with optimized optics can further suppress fluorescence, allowing strong signals from proteins, lipids, and nucleic acids. However, such systems are expensive (often $120,000–$800,000 USD) and require finely tuned configurations, limiting widespread deployment.

Third, the computational demands of spectral unmixing and quantification present a steep learning curve. Generating high-quality Raman images requires expertise across optics, biology, and data science, including custom algorithms for denoising, deconvolution, and quantification. This multidisciplinary barrier restricts high-level Raman imaging to a few expert groups worldwide. For example, the Stevens Lab has pioneered quantitative label-free Raman profiling in tissue engineering, while companies like SuperVision Medicine (China) offer professional Raman-based analysis services for biomedical research.

Despite these challenges, Raman imaging holds great promise. Its ability to perform super-multiplexed, label-free imaging without dyes or destructive processing makes it ideal for spatial metabolomics, early diagnostics, and label-free tissue classification. Emerging fiber-optic and miniaturized Raman systems are being developed for intraoperative use, enabling real-time tumor margin assessment. Ongoing efforts to standardize acquisition, build spectral libraries, and automate data analysis will enhance reproducibility and usability. Ultimately, Raman imaging is poised to become a powerful, non-destructive platform for personalized medicine, drug development, and clinical diagnostics.

Acknowledgements

I would like to sincerely thank all the shareholders of SuperVision Medicine Co. Ltd., China, for their unwavering financial support, which has enabled the continuation and expansion of my Raman-related research since 2022. I am especially grateful to Dr. Vernon LaLone, who first introduced me to the world of Raman imaging during my undergraduate studies at Imperial College London. His passion for and deep love of Raman microscopy left a lasting impression on me. He also inspired me to think boldly, to explore uncharted research territories, and to believe that true success is rooted in love and passion for one's work. I also extend my deepest thanks to Professor Molly Stevens, whose mentorship has profoundly shaped my scientific journey. As a role model to countless young scientists and entrepreneurs, especially women in STEM, her confidence, vision, and relentless pursuit of excellence continue to influence me long after graduation. Her encouragement gave me the confidence to pursue my own independent research path. Finally, I am grateful to Professor Wei Min from Columbia University, whose mentorship and technical insight were instrumental in the founding of SuperVision Medicine. His guidance in the physics and hardware design of high-performance Raman microscopy systems, along with his belief in the importance of commercializing scientific innovation, played a crucial role in the rapid growth of my company. Without his support, this journey would not have been possible.

DOI: http://dx.doi.org/10.5772/intechopen.1012300

Author details

Xiaoyu Zhao
SuperVision Medicine Co. Ltd, Wuxi, China

*Address all correspondence to: xiaoyuzhao@supervision.tech; katherinezxy@qq.com

References

[1] Dillard RS, Hampton CM, Strauss JD, Ke Z, Altomara D, Guerrero-Ferreira RC, et al. Biological applications at the cutting edge of cryo-electron microscopy. Microscopy and Microanalysis. 2018;**24**(4):406-419. DOI: 10.1017/S1431927618012382

[2] Alexandrov T. Spatial metabolomics and imaging mass spectrometry in the age of artificial intelligence. Annual Review of Biomedical Data Science. 2020;**3**:61-87. DOI: 10.1146/annurev-biodatasci-011420-031537

[3] Lichtman JW, Conchello JA. Fluorescence microscopy. Nature Methods. 2005;**2**(12):910-919. DOI: 10.1038/nmeth817

[4] Power RM, Huisken J. A guide to light-sheet fluorescence microscopy for multiscale imaging. Nature Methods. 2017;**14**(4):360-373. DOI: 10.1038/nmeth.4224

[5] Chen F, Tillberg PW, Boyden ES. Expansion microscopy. Science. 2015;**347**(6221):543-548. DOI: 10.1126/science.1260088

[6] Baker MJ, Hussain SR, Lovergne L, Untereiner V, Hughes C, Lukaszewski RA, et al. Developing and understanding biofluid vibrational spectroscopy: A critical review. Chemical Society Reviews. 2016;**45**(7):1803-1818. DOI: 10.1039/C5CS00585A

[7] LaLone V, Aizenshtadt A, Goertz J, Skottvoll FS, Mota MB, You J, et al. Cell-surface Raman nanosensors for label-free analysis. Cell Reports Methods. 2023;**3**:100440. DOI: 10.1016/j.crmeth.2023.100440

[8] Barros da Silva P, Zhao X, Bidarra SJ, Nascimento DS, LaLone V, Lourenço BN, et al. Tunable hybrid hydrogels of alginate and cell-derived dECM to study the impact of matrix alterations on epithelial-to-mesenchymal transition. Advanced Healthcare Materials [Internet]. Nov 2024;**13**(29):e2401032. DOI: 10.1002/adhm.202401032

[9] Zhao X, Wu Y, Kong Z, Wang W, Wang H, Cheng H, et al. Super-multiplexed ultra-quantitative Raman imaging for 3D characterisation of biological sample. Proceedings of SPIE. 2024;**12855**:128550A. DOI: 10.1117/12.3001276

[10] Freudiger CW, Min W, Saar BG, Lu S, Holtom GR, He C, et al. Label-free biomedical imaging with high sensitivity by stimulated Raman scattering microscopy. Science. 2008;**322**(5909):1857-1861. DOI: 10.1126/science.1165758

[11] Zavaleta CL, Smith BR, Walton I, Doering W, Davis G, Shojaei B, et al. Multiplexed imaging of surface-enhanced Raman scattering nanotags in living mice using noninvasive Raman spectroscopy. Proceedings of the National Academy of Sciences of the United States of America. 2009;**106**(32):13511-13516. DOI: 10.1073/pnas.0813327106

[12] Blum C, Schmid T, Opilik L, Weidmann S, Fagerer SR, Zenobi R. Understanding tip-enhanced Raman spectra of biological molecules: A combined Raman, SERS and TERS study. Journal of Raman Specroscopy. 2012;**43**(12):1895-1904. DOI: 10.1002/jrs.4099

[13] Mizushima K, Kumamoto Y, Tamura S, Yamanaka M, Mochizuki K, Li M, et al. Raman microscopy of cryofixed biological specimens for high-resolution

and high-sensitivity chemical imaging. Science Advances. 2024;**10**(50):eadn0110. DOI: 10.1126/sciadv.adn0110

[14] LaLone V, Smith D, Diaz-Espinosa J, Rosania GR. Quantitative Raman chemical imaging of intracellular drug-membrane aggregates and small molecule drug precipitates in cytoplasmic organelles. Advanced Drug Delivery Reviews. 2023;**202**:115107. DOI: 10.1016/j.addr.2023.115107

[15] Herzenberg LA, Parks D, Sahaf B, Perez O, Roederer M, Herzenberg LA. The history and future of the fluorescence-activated cell sorter and flow cytometry: A view from Stanford. Clinical Chemistry. 2002;**48**(10):1819-1827. DOI: 10.1093/clinchem/48.10.1819

[16] Cossarizza A, Chang HD, Radbruch A, Akdis M, Andrä I, Annunziato F, et al. Guidelines for the use of flow cytometry and cell sorting in immunological studies. European Journal of Immunology. 2017;**47**(10):1584-1797. DOI: 10.1002/eji.201646632

[17] Goltsev Y, Samusik N, Kennedy-Darling J, Bhate S, Hale M, Vazquez G, et al. Deep profiling of mouse splenic architecture with CODEX multiplexed imaging. Cell. 2018;**174**(4):968-981.e15. DOI: 10.1016/j.cell.2018.07.010

[18] Krafft C, Schie IW, Meyer T, Schmitt M, Popp J. Raman and CARS microspectroscopy of cells and tissues. The Analyst. 2012;**137**(19):4392-4402. DOI: 10.1039/c2an35181a

[19] LaLone V, Fawaz M, Mourão M, Bradley P, Sisson TH, Shedden K, et al. Bioanalytical Raman spectral cytometry of alveolar macrophages for the quantitative study of idiopathic pulmonary fibrosis. American Journal of Respiratory and Critical Care Medicine. 2019;**199**:A6081. DOI: 10.1164/ajrccm-conference.2019.199.1_MeetingAbstracts.A6081

Pascut FC, Goh HT, Welch N, Buttery LD, Denning C, Notingher I, et al. Noninvasive detection and imaging of molecular markers in live cardiomyocytes derived from human embryonic stem cells [Internet]. Biophysical Journal. 5 Jan 2011;**100**(1):251-259. Available from: https://www.ncbi.nlm.nih.gov/pmc/articles/PMC3010010/

Chapter 7

Multiple Raman Coherent Process in Generation of Biomolecular Complexes

Nicolae A. Enaki, Diana Podoleanu, Ion Munteanu, Gheorghe Chirman, Elena Starodub and Tatiana Paslari

Abstract

This chapter introduces an optical framework for the generation of molecular superstructures from tertiary components through Raman-induced interactions with a bimodal cavity field. A nonlinear model is developed based on cooperative processes analogous to multiple Raman scattering and their integration with biomolecular assemblies. Particular attention is given to phase transitions leading to the formation of supramolecular structures, where induced Raman excitation enables the stabilization of multimeric states such as dimers, trimers, tetramers, and pentamers. The proposed master equation describes the synthesis and stabilization of such structures under Raman excitation. Importantly, the model can also be applied inversely, where large molecular aggregates are dissociated into smaller components *via* induced multiple Raman scattering. This dual mechanism provides a theoretical basis for designing highly efficient optical metamaterials with potential applications in advanced decontamination technologies. To illustrate the applicability, experimental investigations were performed using yeast fungi as a surrogate for more sensitive pathogens (viruses and bacteria). The results revealed two distinct decontamination regimes under nanosecond-pulsed ultraviolet C radiation, consistent with the predictions of the multiple Raman dimerization model. These findings highlight the potential of Raman-driven cooperative processes in both molecular synthesis and selective pathogen inactivation.

Keywords: Raman cooperative effects, biomolecular complexes, multiple Raman scattering, supramolecular synthesis, multimer formation

1. Introduction

The coherent excitation of higher-order Raman scattering in a bimodal electromagnetic field has applications not only in nonlinear optics [1–7], but also in the generation and control of supramolecular structures under external laser irradiation. In this context, Raman excitation induces dipole moments in molecular substructures, providing critical insights into protein bonding. The selective assembly of new molecular aggregates is particularly relevant for drug development, therapeutic applications, and bioengineering.

Moreover, controlling the secondary structures of protein supermolecules is essential for understanding protein folding and cellular function [8–15]. In our previous publication [16–19], we presented new field characteristics of bimodes in a coherent state, in which multistep Raman generation is described by a superposition, $\mathbf{E}^{(+)}(z,t) = \mathbf{E}_p^{(+)}(z,t) + \mathbf{E}_{as1}^{(+)}(z,t) + \mathbf{E}_{as2}^{(+)}(z,t) + \ldots + \mathbf{E}_{asm}^{(+)}(z,t)$, with frequencies of the pump, first anti-Stokes, second anti-Stokes, ..., and m anti-Stokes ω_p, $\omega_p + \omega_r$, ..., $\omega_p + m\omega_r$. Here, $2\hbar\omega_0$ is the excitation energy of each radiator (atom, molecule, exciton). The new characteristic of bimodal excitation of the cavity field was introduced in the collinear cavity/fiber approximation, $\hat{\Pi}^+(t) = \hat{\mathbf{E}}^{(+)}(z,t)\mathbf{E}^{(-)}(z,t)$, which permits us to represent the total intensity correlation as a sum of each component:. This field is described by Raman excitation, $\hat{\Pi}_{1m}^+(t) \sim \hat{E}_{asm}^{(+)}(t)\hat{E}_{asm-1}^{(-)}(t)$; $\hat{\Pi}_{2m}^+(t) \sim \hat{E}_{asm}^{(+)}(t)\hat{E}_{asm-2}^{(-)}(t)$; and de-excitation: $\hat{\Pi}_{1j}^-(t)$, $\hat{\Pi}_{2j}^-(t)$, $\hat{\Pi}_{3j}^-(t)$, components of the bimodal cavity field with frequency differences multiples of the excitation energy, $\hbar\omega_r$: ω_r; $2\omega_r$; $3\omega_r$ respectively. It is demonstrated that the superposition of new components may be used as a guide field characteristic with good phase and amplitude in the semi-classical approximation $\left\langle \hat{\Pi}_{am}^+(t) \right\rangle = P_{km0} \exp[-i\phi_\alpha]$. Here, the Stokes component of generation of bimodal $\hat{E}_j^+(t)$ can be regarded as a pump field for $j - \alpha$ order of Raman Lasing, so that the products like $\hat{E}_j^+(t)\hat{E}_{j-\alpha}^-(t)$ have the same phase, $\phi_{k\alpha} = i\alpha\omega_r t + i\alpha\delta k$, in which $\hbar\alpha\omega_r = \hbar\omega_{j+\alpha} - \hbar\omega_j$ is the portion of excitation energies obtained when the submolecular system passed from ground into the excited bound state in the α-multiple Raman steps ($\alpha = 1,2,3, \ldots$). Here, δk is the wave vector of bimodal components of induced lasing in the cavity. Recent advances in coherent scattering microscopy have enabled the analysis of correlations between anti-Stokes and Stokes components in multiple Raman scattering, using the coherent components stimulated by pump photons [20–28]. The radiation formed by well-correlated bimodes is characterized primarily by the intensity distribution of each field mode. An important aspect of this problem lies in the selective two-quantum excitation of molecular subsystems, where the dipole-active emission must be minimized relative to Raman excitation (see Section 1.2). This principle can be extended to microbiology [29–33], where selective deactivation of molecular structures—such as viral components—may be achievable through induced Raman processes. Consequently, a rigorous description of both amplitude and phase of the radiation generated by bimodal correlated photons becomes essential.

Protein supramolecular clusters play critical roles in processes such as information transfer, cell division, intracellular trafficking, and signal transduction. Their behavior is strongly influenced by structural modifications in proteins and DNA induced by external factors, including radiation and electrical fields.

Section 1.2 examines how biomolecular structures can be reorganized under pulsed radiation through multiple Raman excitation, leading to the synthesis of larger supramolecular assemblies from molecular subunits [34–36]. Radiation formed by the well-correlated bimodes [16–19] is characterized by the intensity of each electromagnetic field component within the bimodal superposition. The generation of such bimodal radiation in cavities, as described in Refs. [16–19], is applied here for the cooperative excitation of bound biomolecular complexes. This interaction is modeled by Hamiltonian (1), which incorporates both the bimodal field and dipole-dipole interactions among tertiary components.

Figures 1 and **2** illustrate the possible interactions of a bimodal field with molecular dipoles located inside cavities or within the evanescent field of fibers or

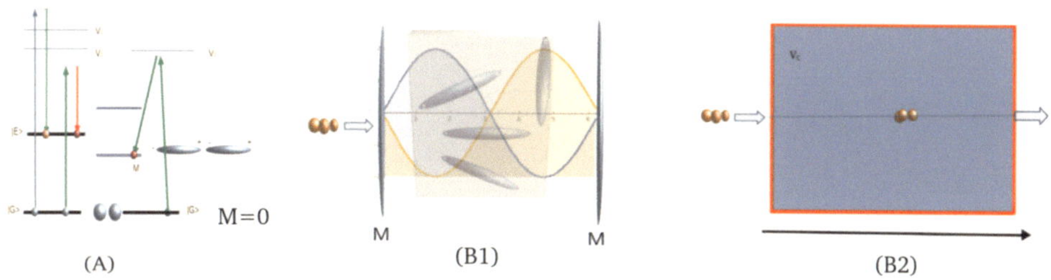

Figure 1.
Generation of molecular substructures with large dipole moments in the excited states: (A) Multiple stages of Raman excitation of the Kasha molecular cluster [14, 15, 37], consisting of two dipoles: (B1) and (B2) possible configurations of such excitation within the cavity. Panel B illustrates the increased dimensions of the molecular dipoles relative to the standing wave.

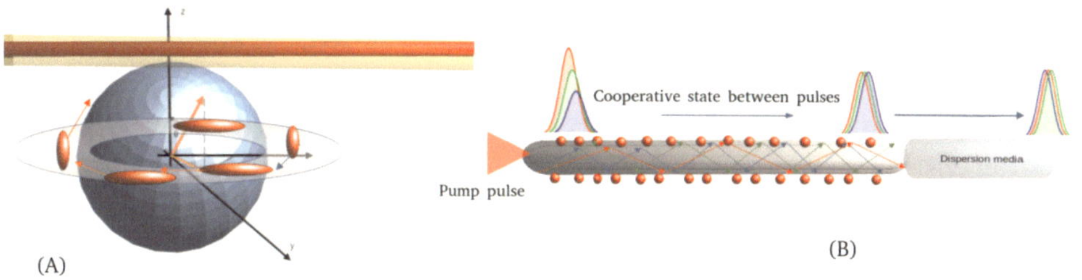

Figure 2.
Interaction of a multi-step bimodal field with submolecular structures in the evanescent field of a microsphere excited via *optical fiber contact (Panel A); multiple Raman excitation of bimolecular dipoles within the evanescent field of an optical fiber (Panel B). The dipoles are shown enlarged to illustrate potential dipole–dipole interactions between the molecular components.*

microspheres. These configurations also highlight potential ways to manipulate dipoles at the nodes or antinodes of the standing wave, as described by the master Eqs. (6), (8), and (12). An attractive feature of this approach is the selective two-quantum excitation of molecular subsystems during multiple Raman processes, which can generate large ensembles of excited dipoles through cooperative interactions between radiation and biomolecules. Solutions of the stationary Schrödinger Eq. (10) further indicate the existence of stable excited states in trimeric, tetrameric, and pentameric configurations.

An alternative perspective arises when selective Raman excitation of DNA disrupts dipole hydrogen bonds [29, 32], creating conditions for bond inactivation and the simultaneous formation of dimeric structures. Such dimerization is the well-established mechanism in pathogen inactivation by UV-C radiation [29, 31].

In this context, the concept of biomolecular synthesis discussed in Section 1.2 can be extended to processes that act in the opposite direction, as described in Section 1.3. To capture these molecular dynamics, we describe the amplitude-phase correlations of radiation regenerated by molecular complexes using the master Eqs. (6), (8), and (12). Developed under the good-resonator approximation, these equations allow for the identification of quantum correlations between non-adjacent Raman orders in bimodal field generation, as previously reported [16–19].

Based on this framework, we introduce the notion of coherent activation of the bimodal field, which governs both the synthesis and inactivation of biomolecular complexes. A numerical example of this process is illustrated in **Figure 3**.

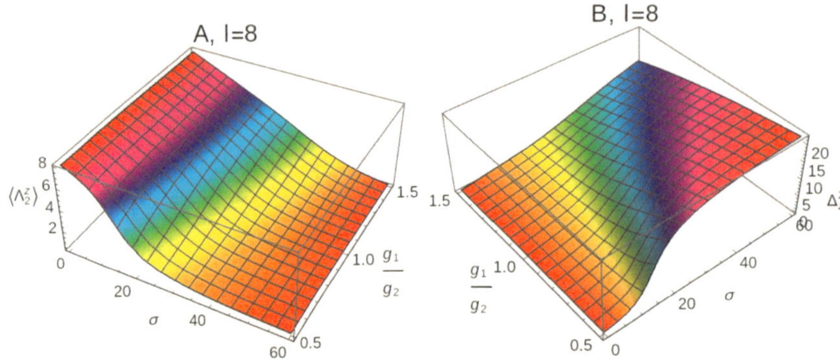

Figure 3.
A. The system of biomolecules is trapped into the coherent state near the excited state, l = 8. This coherent system converts, l = 8, photons into the higher energy quanta of the discrete cavity states. Achieving the cooperative expectation, the conversion process relaxes into the state with inversion near the excited state, $\left\langle \hat{\Lambda}_2^z \right\rangle - l$.

2. Modification of biomolecular structures under multiple Raman coherent excitation

In this section, we discuss the generation of protein structures under multiple Raman coherent excitation of tertiary substructures. Special attention is given to protein sequences, which indicates that information about a protein's native substructure is encoded in its amino acid sequence. We further examine the secondary bonds between tertiary substructures. It should be noted that the problem of extracting this information from the sequence to reconstruct tertiary structures remains unresolved, even for well-studied proteins such as microtubules. Recent investigations, outlined step by step in Refs. [34, 36, 38–42], have focused on identifying the rules by which a primary amino acid sequence folds into secondary structures, how these elements (β-strands and α-helices) assemble into supersecondary structures, and finally, how the architecture of protein assemblies formed from secondary structures can be studied [43–45]. It is of interest to describe these superstructures in reverse order, highlighting how the topology of supramolecular assemblies may influence the formation of the primary structure. The effect of the primary structure on secondary structure formation was successfully demonstrated in early studies following the topological analysis of horizontal protein structures [36, 38–41]. However, this predictive capability diminishes in later stages, where a novel description of protein substructure packing is required. In this context, local symmetry and concepts associated with the propagation of collective excitations along the protein may play a decisive role [43–45].

UVC radiation has demonstrated its potential in eliminating pathogenic microorganisms through complex physicochemical mechanisms and has recently attracted significant attention for biomedical applications, including surface decontamination and rapid diagnostics [31, 46]. To date, no predictive model exists for the primary modification of biomolecular structures under radiation. This section describes protein molecular modifications, as well as DNA/RNA dimerization, stimulated by UVC radiation during Raman interactions with secondary bonds in supramolecular structures [47–56]. The proposed model accounts for modifications of protein supramolecular structures or DNA/RNA dimerization as a function of applied pulsed UVC radiation. Statistical detection of multiple-photon conversion processes may facilitate the generation or modification of supramolecular structures through dimerization, forming new secondary bonds between

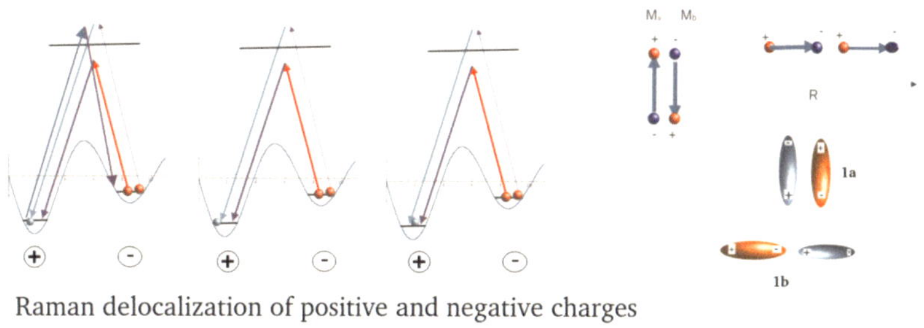

Figure 4.
Spatial delocalization of positive charges relative to negative ones creates long-lived dipoles after Raman excitation of each radical. In the case of the two structures of polypeptide protein subsisted, a = I and b = II the two molecular chains of amino acids are involved in dipolar interactions, as shown in ***Figure 1a, b*** *for vertical antiparallel (a) and parallel horizontal arrangement head to tail (b).*

DNA and proteins, based on cooperative processes analogous to Raman multiple scattering. The model further predicts changes in the fluorescence spectra of biomolecules corresponding to modifications in dipolar secondary structures under pulsed UVC radiation. This theory suggests that biomolecular dimerization is strongly influenced by UVC light intensity, indicating a quantum mechanism of interaction with DNA and proteins, particularly affecting tertiary bonds, as illustrated in **Figure 4**.

For the description of the intensity dependence of the modification of biomolecular structures under the action of pulse radiation, we propose the Born-Oppenheimer (BO) approximation, frequently used in chemistry and microbiology. In this approach, the nuclei subsystem is characterized by the vectors, $\mathbf{R} = \{R_1, R_2, \ldots, R_n\}$, and is considered a slow subsystem compared to the electronic one, $\mathbf{r} = \{r_1, r_2, \ldots, r_n\}$. The density matrix of the system is represented through the wave function of molecular complex, $\rho(t, \mathbf{r}, \mathbf{R}) = |\Psi_e(t, \mathbf{r}, \mathbf{R})\rangle|\chi(\mathbf{R})\rangle\langle\Psi_e(t, \mathbf{r}, \mathbf{R})|\langle\chi(\mathbf{R})|$. This allows us to represent the molecular wave function by the product of the wave functions, $|\Psi_e(\mathbf{r}, \mathbf{R})\rangle|\chi(\mathbf{R})\rangle$, of the subsystems of electrons, $|\Psi_e(\mathbf{r}, \mathbf{R})\rangle$, and the subsystem of nuclei, $\chi(\mathbf{R})\rangle$, of the respective molecules. The modification of biomolecular structures like proteins takes place under the action of UVC radiation, as schematically represented in **Figure 5**. The similar modification of hydrogen bonds into covalent ones under the Raman processes is represented in **Figure 6** described in the literature as a modification of interaction potential during the dimerization. Considering that the nuclei positions are fixed for the electronic subsystem, the stationary Schrodinger equation can be represented in the form, $H_e(\mathbf{r}, \mathbf{R}_s)|\Psi_e(\mathbf{r}, \mathbf{R}_s)\rangle = E_e(\mathbf{R}_s)|\Psi_e(\mathbf{r}, \mathbf{R}_s)\rangle$. Obviously, the electronic energy eigenvalue E_e depends on the chosen positions $\mathbf{R}$ of

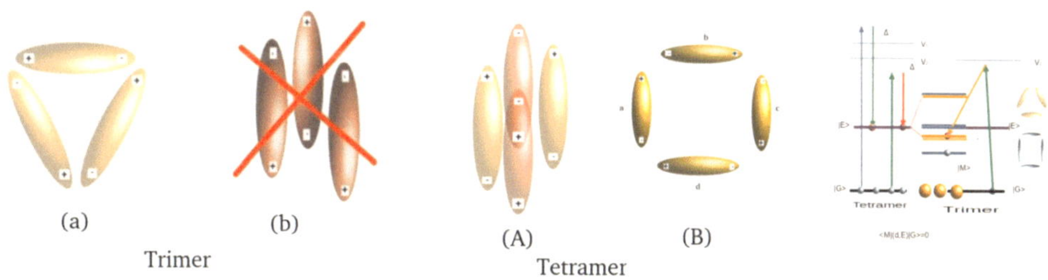

Figure 5.
Representation of a trimer protein (a and b) and tetramer ones (A and B) through the dipole–dipole interaction in the possible H and J dipole configurations of molecular aggregates. Formation of a trimer is not possible in the vertical dipolar bond. Panel C shows the spectrum obtained from the solution of the stationary Schrödinger equation for a single cooperative excitation in the trimer and tetramer.

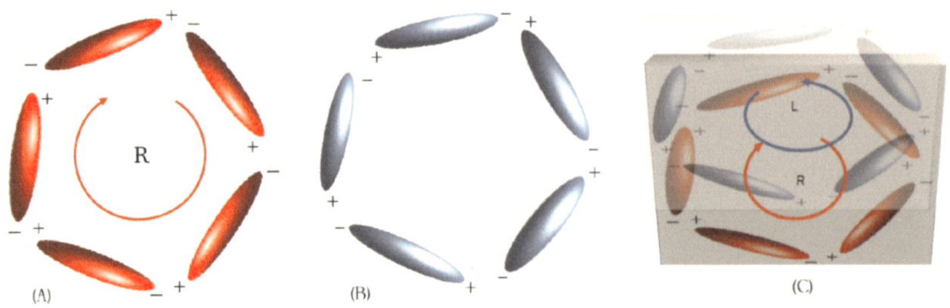

Figure 6.
The formed multimers in the right polarization (A) may induce the left polarization packing (B) in the new pentamer biomolecule, attracting a new structure, like that represented in Fig. (C).

the nuclei. Varying these positions R in small steps and repeatedly solving the electronic Schrodinger equation, one obtains $E_e(\mathbf{R}_s)$ as a function of $\mathbf{R}_s$. As is known, the excitation of bi-molecular radical can substantially delocalize the positive and negative charges as this is represented in 3), so that this effect creates an additive dipole-dipole interaction between the biomolecular subsystems [33]. In this adiabatic description, the master equation also may be factorized, $\rho(t, \mathbf{r}, \mathbf{R}) = \hat{W}(t) \otimes \hat{\rho}_N(\mathbf{R}_0)$, where $\hat{W}(t) = |\Psi_e(t, \mathbf{r}, \mathbf{R_0})\rangle\langle\Psi_e(t, \mathbf{r}, \mathbf{R_0})|$ is the excited electron subsystem in each submolecular cluster and $\hat{\rho}_N(\mathbf{R}_0) = |\chi(\mathbf{R})\rangle \langle\Psi_e(t, \mathbf{r}, \mathbf{R})\rangle|\chi(\mathbf{R})|$ is the density matrix of nuclei subsystem which is slowly changed after the reconfiguration of the electronic subsystem in each cluster.

Let us now discuss the possible modifications in the electron subsystem when the nuclei are fixed. According to perturbation theory, the transition matrix element in each step is proportional to the product of dipole active transition elements from the ground and excited state of Kasha submolecular components to the nearest virtual one and inversely proportional to the detuning from resonances, with the virtual state, $\Delta_p, g_p \sim d_{ev_p} d_{v_p g} q_{p-1} q_p / (\hbar^2 \Delta_p)$ (see **Figure 1a** and **4**). Here, d_{ei} and d_{ig} are the dipole transition matrix elements from the excited/ground state to the intermediary state, $|v_p\rangle$.

q_{p-1} and q_p are the per photon strength of the two adjacent cavity modes, $p-1$ and p of the cavity. The Hamiltonian of such a two-level system connecting a bimodal cavity field in the higher-order Raman scattering process can be represented through the free, $\hat{H}_{C0}$, and interaction, $\hat{H}_I$, parts by the expressions

$$\hat{H}_C = \hat{H}_{C0} + \hat{H}_I + \hat{H}_{IM}, \tag{1}$$

$$\hat{H}_{C0} = \hbar \sum_{p=0}^{n} \omega_p \hat{c}_p^\dagger \hat{c}_p + \hbar\omega_r \hat{D}_z, \tag{2}$$

$$\hat{H}_I = \hbar\chi_n \left\{ \hat{\Lambda}_n^- \hat{D}^+ + \hat{\Lambda}_n^+ \hat{D}^- \right\}, \tag{3}$$

$$\hat{H}_{IM} = \sum_{l,j=1}^{N_r} \chi_{j,l} \hat{D}_j^+ \hat{D}_l^- . \tag{4}$$

Here, we represent the annihilation, $\hat{\Lambda}_n^- = \sum_{p=0}^{n-1} g_{p+1} \hat{c}_p \hat{c}_{p+1}^\dagger / \chi_n$, and creation, $\hat{\Lambda}_n^+ = \sum_{p=0}^{n-1} g_{p+1} \hat{c}_{p+1} \hat{c}_p^\dagger / \chi_n$, operators in the higher-order processes of Raman lasing, where the scattering process takes place from the ground state to the excited state of each dipole component represented in **Figures 1** and **2** system; $\hat{c}_n\left(\hat{c}_n^\dagger\right)$ and $\hat{c}_p\left(\hat{c}_p^\dagger\right)$, correspond to the pump and p-order Stokes modes generated in the $n-$ order Raman-induced lasing in the cavity with the decrease of the photon energy. These operators satisfy the Bose commutation rules: $\left[\hat{c}_{p'},\hat{c}_p^\dagger\right]=\delta_{j,i}$, and $\left[\hat{c}_{p'},\hat{c}_p\right]=0, p',p\equiv 0,1,2,\ \dots,n$. The lowering, $\hat{D}^-$, and excitation, $\hat{D}^\dagger$, sub-molecular (or vibration) operators are connected with the inversion, $\hat{D}_z$ relative to the ground state through the commutation relation, $\left[\hat{D}^+,\hat{D}^-\right]=2\hat{D}_z$, where, $\hat{D}^+=\hat{D}^+=\sum_{j=1}^{N_r}\hat{D}_j^+$ ($\hat{D}^-=\sum_{l=1}^{N_r}\hat{D}_l^-$), and $\hat{D}_z=\sum_{j=1}^{N_r}\hat{D}_{zj}$, are the excitation (lowering) and inversion operators of this submolecular ensemble of generated active dipoles represented in **Figures 1** and **4**. Here, $\hat{D}_i^+=|e_i\rangle\langle g_i|$ and $\hat{D}_i^-=|g_i\rangle\langle e_i|$ are the excited and lowering operators of each submolecular dipole. The submolecular components may enter in the collective excitation operator depending on the number of the dipole active radicals, N_r: $\hat{D}^\dagger=\hat{d}^\dagger\sqrt{N_r-\hat{d}^\dagger\hat{d}}$ and $\hat{D}^-=\sqrt{N_r-\hat{d}^\dagger\hat{d}}\hat{d}$. For big number of radicals in the ground state with small electric dipoles, $N_r\gg\left\langle\hat{d}^\dagger\hat{d}\right\rangle$ may be regarded as boson creation and annihilation operators, $\hat{D}^\dagger\simeq\hat{d}^\dagger\sqrt{N_r}$, $\hat{D}^-\simeq\sqrt{N_r}\hat{d}$.

The free part of the Hamiltonian contains the field energy of each bimodal component of the cavity field represented in **Figures 1** and **2**, as well as the energy of the excited permanent dipoles described by the inversion, $\hat{D}_z$, and the last term (4) of the Hamiltonian describes the interaction between submolecular components. This interaction may be like dipole-dipole interaction between neighboring components as this is represented in **Figures 5** and 7,

$$\hat{H}_{IM}=\sum_{l\in(a),j\in(b)}\left\{\frac{\left(\hat{\mathbf{M}}_j,\hat{\mathbf{M}}_l\right)}{R^3}-3\frac{\left(\hat{\mathbf{M}}_j,\mathbf{r}\right)\left(\hat{\mathbf{M}}_l,\mathbf{r}\right)}{R^5}\right\}, \tag{5}$$

where $\hat{\mathbf{M}}_j$ and $\hat{\mathbf{M}}_l$ correspond to the operator of generated dipoles in substructures (a) and (b) during the delocalization of the positive and negative charges as this is represented in **Figure 5**. This model may be applied to the DNA strands as represented by **Figure 8a,b**, always to extend it on the explanation of the forming of new phosphodiester bonds (p) at the hydroxyl group of the nucleotide. For example, the DNA sequence $GAATTC$ is actually $50-pGpApApTpTpCOH-30$, where adenine, (A), and guanine, (G) are pyrimidines; cytosine, (C), and thymine, (T), are purines [65]. According to Ref. [65], the phosphodiester bonds form the backbone of DNA

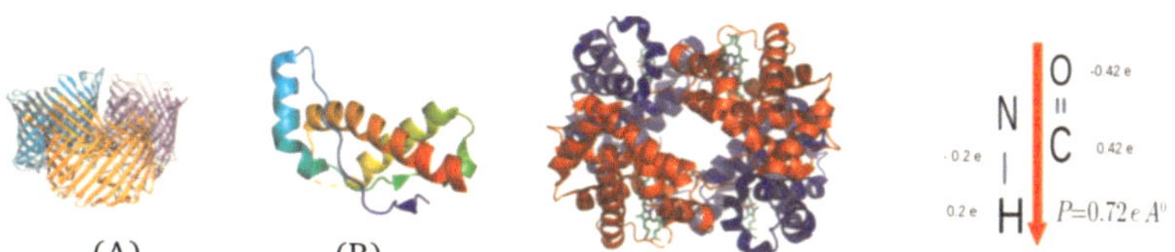

(A) (B) (C)

Figure 7.
Trimer proteins have been observed experimentally in Ref. [34] "A Pandora's Box" and in Ref. [35], where the protein "Prions" is associated with various forms of cerebral encephalitis.

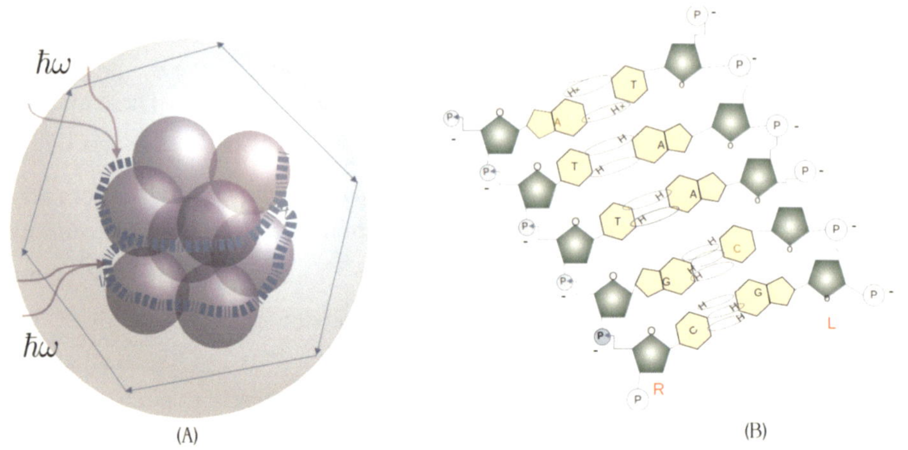

Figure 8.
(A) Observation of spherical histone elements involved in DNA packing [57–60]. Shown is the possible induced localization of the electromagnetic field during Raman interaction with biomolecules (such as histones and DNA) within cell nuclei. (B) Possible dipole–dipole interactions between the right and left DNA strands during transcription, or during inactivation processes associated with dimerization or tautomeric transitions [52, 55, 56, 61–64].

strands and are responsible for the characteristic negative charges on DNA molecules that stabilize DNA double helices as shown in **Figure 8A** and the phosphodiester bonds between the nucleotide may be represented as $\chi_{j,l}$. Here, of course, in the interaction Hamiltonian, the nucleotides are not separated in pyrimidines/purines sequences. We are interested in the Raman excitation of such biomolecules and possible creation of molecular micro-dipoles in interactions, as this is represented in **Figures 4–7**.

A. The behavior of the molecular subsystem. Considering that the field may generate such a bistable state relative to the excitation between unbounded ground and bonded supramolecular states, we propose to generate these supra-structures by a Raman excitation process. Let us discuss the possibilities of the stabilization of the standing wave near the surfaces with a higher refraction index. For example, in many eukaryotic cells, the refractive index is about for phytoplasma/cytosol; for nucleus; 1.375 – 1.385 for nucleolus; 1.400 – 1.420 for mitochondria; 1.600 for lysosome [47–51]. These indexes is relative larger than the refractive index of water, $n_0 = 1.3325$, and intra-cellular fluids is about $n_{0f} = 1.35$ [47, 48].

Both cytoplasm and cytosol are located within the boundaries of the plasma membrane and surround the nucleus in eukaryotic cells. They play crucial roles in the supporting cellular activities by providing a medium for chemical reactions, as illustrated in **Figure 6**. This optical material may serve as an effective internal reflector, directing light waves into the nucleus when the incident angle is below the critical angle defined by Snell's law. This effect can localize a standing wave near the surface of a quasi-spherical cavity, like gallery waves, potentially inducing coherent Raman excitation of the bound molecular structures described above *via* the multistep scattering term of Hamiltonian (3). We propose a model of cooperative excitation describing the coupling of tertiary protein subsystems, as shown in **Figure 5a,b** for trimers and **Figure 5A,B** for tetramers.

On the bases of mater equations obtained in the bad cavity limits [63], and in the good cavity approximation described by Refs. [17–19], we introduce the master equation for the description of the statistical aspects of generation or dissociation of proposed molecular structures in **Figures 4–8**. Denoting by the excitation, $\hat{R}_i^+$, and lowering, $\hat{R}_i^-$, operators in right direction of the chain, and by the similar operators the dipole positive defined direction of the chain of the dipoles constructed by tile to cap connections, the chain detected in opposite sense, $\hat{L}_i^+$ and $\hat{L}_i^-$, we may define the Hamiltonian for the creation of number of protein molecules under the action of bimodal cavity field (3). Following the method proposed in Ref [63], we consider the bad cavity limits when the two-quanta Rabi frequency is less than the cavity losses. In this approximation, we regard the bimodal cavity field as a coherent field for su(2) algebra in interaction with a two-level system that readily represents the bonded and unbonded molecular dipole components described by the master equation,

$$\frac{\partial}{\partial t}\hat{W}(t) = \frac{i}{\hbar}\left[\hat{W}(t), \hat{H}_I\right] + \hat{L}\hat{W}(t), \tag{6}$$

where the Hamiltonian part of master equation is,

$$\begin{aligned}\hat{H}_I = {} & \hbar\Delta_r\sum_i \hat{R}_i^z + \hbar\Delta_l\sum_i \hat{L}_i^z + \\ & -\hbar\sum_{i,j}\chi_{ij}^{(r)}\hat{R}_i^+\hat{R}_j^- - \hbar\sum\chi_{ij}^{(l)}\hat{L}_i^+\hat{L}_j^- - \hbar\sum_{\langle i,j\rangle}\chi_{ij}^{(lr)}\left(\hat{L}_i^+\hat{R}_j^- + \hat{R}_j^+\hat{L}_i^-\right) \\ & +\hbar\Omega_r\sum_i \sin(\delta k R_i)\left(\hat{R}_i^+ + \hat{R}_i^-\right) + \hbar\Omega_l\sum \sin(\delta k R_i)_i\left(\hat{L}_i^+ + \hat{L}_i^-\right),\end{aligned} \tag{7}$$

In master Eq. (6), the terms $\hat{L}\hat{W}(t)$ describe the losses from the cavity. Here, the interaction between the neneighborin protein subsystem takes place, $\sum_l \chi_{ij}^{(r)}\hat{R}_i^+\hat{R}_j^- = \chi_{ii-1}^{(r)}\hat{R}_i^+\hat{R}_{i-1}^- + \chi_{ii+1}^{(r)}\hat{R}_i^+\hat{R}_{i+1}^-$. The similar sum is represented and for the left dipole circuit, $\sum_j \chi_{ij}^{(l)}\hat{L}_i^+\hat{L}_j^- = \chi_{ii-1}^{(l)}\hat{L}_i^+\hat{L}_{i-1}^- + \chi_{ii+1}^{(l)}\hat{L}_i^+\hat{L}_{i+1}^-$; $\Omega_r(R_i) = \kappa\sum_{j=1}^k g_k\langle\hat{\Lambda}_k^-\rangle_r \sin(\delta k R_i)$, and $\Omega_l = \kappa\sum_{j=1}^k g_k\langle\hat{\Lambda}_k^-\rangle_l \sin(\delta k R_i)$ are the Rabi frequencies of the standing wave of the bimodal cavity field represented in **Figure 1**, and $\delta k = k_{i+1} - k_i$ is the difference between the pump, k_p and Stokes wavelength, k_p, in the case of single-step Raman excitation of biomolecular complexes. Here, we represent the R- and L rings/filaments packing configuration represented in **Figure 6b**. The interactions with each other of molecular rings/filament components are described by the expression; $\sum_{j=i\pm1}\chi_{ij}^{(lr)}\hat{L}_i\hat{R}_j^-$; $\Delta_r = \omega_r - \omega$; $\Delta_l = \omega_l - \omega$, where ω_r, and ω_l are the excitation frequencies of the right and left protein filaments represented in **Figures 5**, **8**, and **9**; $\omega = \omega_{i+1} - \omega_i$ is the difference between the pump and Stokes components in each step of Raman excitation (in the case of a single step, we have $\omega = \omega_p - \omega_s$). As the dipole-dipole interaction (5) at large distance may be less than Raman coherent excitation of sub-molecular dipoles we neglect this term from Hamiltonian (1), considering that $1/\delta k$ is smaller than the main distance between the dipoles enlarged represented in the standing wave of bimodal cavity field of **Figure 1**, B1, and described by inequality $M_iM_j1/R^3 \ll \Omega_{r0}\sin(\delta k R)$. Here, $\Omega_{r0}$is the amplitude of modulated Rabi frequency in the cavity. As the interaction Hamiltonian with multi-mode coherent Raman field may induce the compression of the dipoles, the compression force of dipoles, $F = -\partial\langle\hat{H}_I\rangle/\partial R_i$, into nodes or ridges of the standing wave shown in

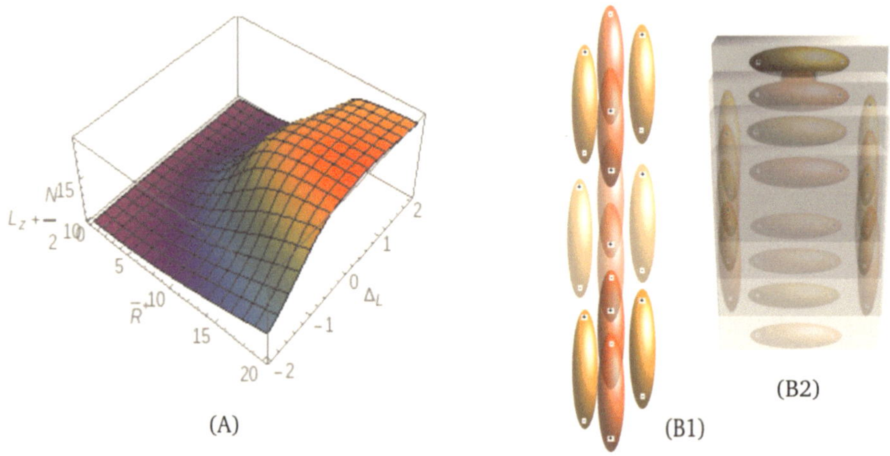

Figure 9.
(A) The induced of new multimer structures within the spatial configuration described in ***Figure 6(C)*** *may be explained by the action of left-handed polarization of a pentamer on the corresponding right-handed one (see solution (8)). Possible packing arrangements of tetramers are shown in H-type (Figure B1) and J-type (Figure B2) three-dimensional structures correspond to filament-like and ring-like organizations, respectively.*

Figure 1 can stimulate dipole-dipole interactions between the components of generated biomolecules.

To describe this effect, we impose the condition that the Raman resonance relative to the excited state of bound molecules satisfies: $\hbar\Delta_r/2 + (E_n - I) = -\hbar\Delta_r/2$; for trimer, tetramer, and pentamer, this condition becomes, $\Delta_r = 2w/\hbar$. Here is the generation of a super molecule relative to the ground state, $E_i - I_1$; the left rings stimulated by right ones are described in **Figures 5** and **8**. Introducing the new renormalized molecular operators, we obtain the master equation which takes into consideration the interaction of two filament/ring structures,

$$\begin{aligned}\frac{\partial}{\partial t}\hat{W}(t) = {} & \left[i\Delta_L \hat{L}_i^z, \hat{W}(t)\right] + [\hat{L}_i^+, \left(\gamma_{lj} - i\chi_{ij}^{(l)}\right)\hat{W}(t)_j^- \\ & + \left(\gamma_{lr} - i\chi^{(lr)}\right)\hat{W}(t)\hat{R}_i^-] + [\hat{R}_i^+, \gamma_r \hat{W}(t)\hat{R}_i^- \\ & + \left(\gamma_{lr} - i\chi^{(lr)}\right)\hat{W}(t)\hat{L}_i^- - i\Omega_r \hat{W}(t)\}] + H.c.\}.\end{aligned} \qquad (8)$$

Here γ_l, γ_r, and γ_{lr} are the photon losses from the cavity during the cooperative excitation of bounded proteins. We require that the density matrix be expressed by the normal product $\hat{W}(t_{stat}) = \hat{F}\left(\hat{R}^+, \hat{L}^+\right)\hat{F}^*\left(\hat{R}^-, \hat{L}^-\right)$. Let us consider that the external bimodal field exits only the first ring from the coupling system represented in **Figures 5** and **8** so that, $\Omega_r \gg \Omega_l$. In this particular case, we obtain the stationary equation only for the right filament/ring represented by the expression,

$$\left[\hat{R}_r^+, \left\{\gamma_r \hat{R}_i^- + \left(\gamma_{lr} - i\chi^{(lr)}\right)\hat{L}_l^- - i\Omega_r\right\}\hat{F}^*\left(\hat{R}^-, \hat{L}^-\right)\right] = 0; \qquad (9)$$

Considering that for an R filament/ring the interaction with L filament, we neglect the term $\left(\gamma_{lr} - i\chi^{(lr)}\right)\hat{L}_l^-$ from the expression (9) in comparison with $i\Omega_r$. In this situation, the solution is reduced to a well-known expression [66–68],

$$\hat{F}^*(\hat{R}^-, \hat{L}^-) = \prod_{l=1}^{N} \frac{1}{-i\Omega_r + \gamma_r \hat{R}_l} f(\hat{L}^-).$$

Here, R denotes the number of right-handed circuits in the possible packing configurations shown in **Figure 6**. In this case, we neglect the interaction between right-handed rings that are separated by left-handed ones. If the right ring molecules possess only two states, then $R = 2$. In this situation, we obtain the identity.

Here, N denotes the number of right-handed circuits in the possible packing configurations shown in **Figure 6**. In this case, we neglect the interaction between right-handed rings that are separated by left-handed ones. If the right ring molecules possess only two states, then $\hat{R}_i^{+2} = 0$. In this situation, we obtain the identity, $1/(-i\Omega_r + \gamma_r \hat{R}_i^-) = \frac{1}{-i\Omega_r} \exp\left[-i\frac{\gamma_r}{\Omega_r}\hat{R}_l^-\right)$. The solution for the R ring becomes, $\hat{W}\left(\hat{R}^+, \hat{R}^-\right) = A \exp\left[+i\frac{\gamma_r}{\Omega_r}\sum_{l=1}^{N}(\hat{R}_l^-)\right] \exp\left[-i\frac{\gamma_r}{\Omega_r}\sum_{m=1}^{N}(\hat{R}_m^-)\right]$. From this expression, it follows that the rings evolve into a coherent state. This coherent state, in turn, induces the same type of coherence in the left dipole–dipole "circuits," where instead of an external field, the polarization of the right circuit acts as the driving mechanism. As a result, a nonzero polarization also appears in the left ring/filament (see **Figure 6**). The solution of this stationary equation is described by the expression,

$$\hat{W}\left(\hat{L}^+, \hat{L}^-\right) = D\left(\frac{1}{\hat{L}^- + \chi^*\left\langle\hat{R}^+\right\rangle}\right)^{\frac{-i\Delta_L}{\left(\gamma + i\overline{\chi}^{(l)}\right)}+1} \left(\frac{1}{\hat{L}^- + \chi\langle\hat{R}^-\rangle}\right)^{\frac{i\Delta_L}{\left(\gamma - i\overline{\chi}^{(l)}\right)}+1},$$

where $\chi = (\gamma lr - i\chi^{(lr)})/(\gamma - i\overline{\chi}^{(l)})$, and D and A are normalization constants.

A similar situation may occur in DNA, where dipole–dipole interactions between base pairs stabilize the system relative to the dipole-active states of symmetric, unbonded nucleobases [47–56, 63–65]. In other words, the dipole–dipole coupling between hydrogen bonds in nucleobases can protect DNA from active dipole excitation by solar radiation, effectively making it resistant to external quantum correlations (see **Figure 8**). Specifically, considering the two opposing dipoles of the hydrogen bonds between adenine and thymine, the degenerate ground states, $(A :: T)$, the degenerate ground states, I_{1a}, and I_{1b}, of the two hydrogen bonds may split into two levels, according to perturbation theory as described by Eq. (10). Denoting these two hydrogen bonds as a and b, we find the relation, $(I_{1a} - E)((I_{2a} - E) - w^2$, where, in the case where the hydrogen bond dipoles are antiparallel and oriented perpendicular to their separation vector, the interaction energy is given by, $W = \langle 1b|\hat{H}_{IM}|1a\rangle \sim -d_a d_b/r^3$. Since the individual bond energies I_{1a} and I_{1b} differ slightly, $I_{1b} = I_{11} + \Delta$, the ground state energies are shifted and split in a manner analogous to that observed for multimers.

Assuming that excitation occurs from a state with a small electric dipole moment (the fundamental state), $|g_1, e_2, \ldots, g_N\rangle$, to an excited state, $|e_1, \ldots, g_{n-1}, g_n\rangle$, where each substructure has a total dipole moment, d_i, described by the secular equation,

$$\left(\hat{H}_0 + \hat{H}_{IM}\right)|E_\alpha\rangle = E_\alpha|E_\alpha\rangle, \quad \alpha = 1, 2, \ldots \tag{10}$$

The transfer of each protein subunit from the ground state to the excited state is taken into account in the multiple Raman model as defined in Refs. [17–19] and can be expressed through the excitation energy of each peptide chain, E_n. For example, the energy of the excited state is determined by three wave vectors, $|e,g,g\rangle$, $|g,e,g\rangle$, $|e,e,g\rangle$, and is calculated relative to the base state, $|e,e,e\rangle$, with energy, I_1. This is represented by the matrix element, $W = -w < 0$, in the form: $E_1^{(1)} = I_1 + w$; $E_1^{(1)} = I_1 + w$, and $E_3^{(1)} = I_1 - 2w$. The corresponding wave functions can be calculated directly. Below we present the analytical expression of the underlying state, $|E_1^{(1)}\rangle = \{-|e_a, g, g_c\rangle + |g_a, g_c, e_b\rangle\}/\sqrt{2}$; $|E_2^{(1)}\rangle = \{-|e_a, g, g_c\rangle + |g_a, e_b, g_c\rangle\}/\sqrt{2}$, and $|E_3^{(1)}\rangle = \{|e_a, g_b, g_c\rangle + |g_a, e_b, g_c\rangle + |g_a, g_b, e_c\rangle\}/\sqrt{3}$. When interacting with the vacuum of the electromagnetic field, this state does not relax into the ground state, $|G\rangle = |g_a, g_b, g_c\rangle$, and the resulting superposition corresponds to a stable molecular configuration. This state represents the proper trimer with a comparatively long lifetime relative to the base state, due to the fact that the matrix element vanishes, $(\langle E_3|\mathbf{d}|G\rangle, \mathbf{E}(t)) = 0$, where, $\mathbf{d} = \sum_{i=a}^{c} \mathbf{d}_i$. If we consider the field oriented along the dipole moment, a, the scalar product vanishes, $E\{d_a^{eg} - d_b^{eg}\cos(\pi/3) - d_c^{eg}\cos(\pi/3)\} = 0$. If the field is oriented along the bisectors of the equilateral triangle, the mean value of the scalar product vanishes too, $E\{0 - d_b^{eg}\cos(\pi/6) - d_c^{eg}\cos(\pi/6)\} = 0$. The second excitation state of trimer molecule can be expressed through the superposition, $|E_\alpha\rangle = B_{\alpha,1}|e_a, e_b, g_c\rangle + B_{\alpha,2}|g_a, e_b, e_c\rangle + B_{\alpha,3}|e_a, g_b, e_c\rangle$. The degenerate energy splits into new states: $E_1^{(2)} = I_2 + w$; $E_2^{(2)} = I_2 + w$, and $E_3^{(2)} = I_2 - 2w$. The wave functions for these states can also be calculated directly. Finally, the analytical expression of the underlying state is presented: $|E_3^{(2)}\rangle = \{|e,g,e\rangle + |e,e,g\rangle + |g,e,e\rangle\}/\sqrt{3}$, and $|E_{1,2}^{(2)}\rangle = \{2|e,g,e\rangle - [|e,e,g\rangle + |g,e,e\rangle]\}/\sqrt{3\times 2}$. In both cases, the spectrum can be diagonalized so that the left- and right-handed circuits correspond to the new eigenvalues.

We may extend the construction of multimer biomolecules to the case of five submolecular dipoles, as represented in **Figure 6**. The eigenvalues of the energy and the corresponding eigenvectors are obtained from the system of Eq. (10), where the wave function is expressed as a superposition, $|E_\alpha^{(1)}\rangle = \sum_{i=1}^{5} C_{\alpha i}$. The zero values of the principal determinant of the system of Eq. (10) give us the energy splitting of the degenerate spectrum of five dipoles represented in **Figure 6** and described by the equation,

$$(I_1 - E_\alpha)^4 + 2w(I_1 - E_\alpha)^3 - w^2(I_1 - E_\alpha)^2 - 2w^3(I_1 - E_\alpha) + w^4 = 0.$$

The solution of this equation is given by the expressions, $E_5 = I_1 - 2w$, $E_{4,3} = I_1 + w(1 - \sqrt{5})/2$, $E_{2,1} = I_1 + w(1 + \sqrt{5})/2$. The wave vector for the ground state is represented by the expression,

$$|\mathrm{E}_5^{(1)}\rangle = \frac{1}{2}\{|e,g,g,g,g\rangle + |g,e,g,g,g\rangle + |g,g,e,g,g\rangle + |g,g,g,e,g\rangle + |g,g,g,g,e\rangle\}. \tag{11}$$

Considering that the external field acts only on the right-hand rings, we represent these rings as being periodically arranged. The connections between the components are not yet formed; the left-hand rings are shown in **Figure 5**. For the pentamer, the corresponding energies are given as: $E_{1,2} - I_1 = w(1 + \sqrt{5})/2$; $E_{1,2} - I_1 = w(1 - \sqrt{5})/2$; $E_{1,2} - I_1 = w(1 + \sqrt{5})/2$, and $E_5 - I_1 = -2w$. The numerical aspect of this model can be generalized to the formation of macromolecules consisting of a larger number of submolecular amino acid tertiary components. The key problem lies in the possibility of generating such subsystems using an external coherent field. It is straightforward to determine the wave functions corresponding to the split energy levels. For example, if the system is prepared in the lower excited state with energy, $E_5 = I_1 - 2w$, the wave function remains a symmetric superposition of the five atoms, in which a single excitation is shared collectively among all subsystems—trimer, tetramer, and pentamer.

B. Raman detection. We are interested in detecting information about molecules generated by an external field during Raman excitation of long-lived biomolecular complexes. This information can be obtained by observing the cooperative conversion of pump photons into Stokes photons. Let us first eliminate the atomic polarization by considering the special preparation of the atoms in an initial coherent state, as proposed in Refs. [16–19]. To do this, we assume that the lifetime of the atomic polarization between the excited and ground states of the radiators is shorter than the inverse of their Rabi oscillation frequency.

Photon conversion occurs between the two cooperative modes of the cavity as depicted in **Figure 1A,B**, and described by the Hamiltonian (7) of the molecular system in interaction with the bimodal cavity field. The possibility of realization of such a cavity in cell nuclei is illustrated in **Figure 8A**. The rate at which atoms transition to the ground state, $\partial\langle D_z(t)\rangle/\partial t$, can be detected outside the cavity using ionization techniques. This detection method provides information about the generation rate of the mean number of energy portions, $\partial\left\langle \hat{W}_{ph}(t)\right\rangle/\partial t$, emitted during the evolution time of the biomolecular components through the optical resonator. Considering that the quasiparticle operators, $\hat{\Lambda}(t)$, as a function of $\hat{\Lambda}_n^-(t)$, $\hat{\Lambda}_n^+(t)$, and $W_{ph}(t) = \hat{\Lambda}_n^z + L/2$, we can express the atomic operators in terms of their free and source components, as proposed in Refs. [17–19]. The conservation laws follow directly from the system Hamiltonian (7). We observe that, due to multiple conversions in the system, the total number photons, $\hat{N}_{ph} = \sum_{p=0}^{n} \hat{c}_p^\dagger \hat{c}_p$, and the total number of molecules in excited and ground states, $N_r = \sum_j \left\{ |e\rangle\langle e|_j + |g\rangle\langle g|_j \right\}$, are conserved. Moreover, in the resonance case, the free part of the Hamiltonian commutes with the interaction one, $\hat{H}_I$, allowing us to introduce the conservation of the total number of excitations of portions of energy in the bimodal field and molecular excitations, $\hat{W}_C(t) = \sum_{m=0}^{n} m\hat{c}_m^\dagger(t)\hat{c}_m(t) + \hat{D}_z + N/2 = Const$. The last conservation low opens the possibility to generate in the cavity energy portions multiple to $\hbar\omega$. If we wish to detect the number of energy portions generated in the bimodal cavity field, we may use the operator $\hat{W}_{ph} = \sum_{m=0}^{n} m\hat{c}_m^\dagger(t)\hat{c}_m(t)$. We can use this operator representation in the free part of Hamiltonian, $H_0 = \hbar\omega\hat{W}_{ph} + \hbar\omega_r\hat{D}_z + \hbar\omega N_{ph}$, and follow the traditional approach of resonant interactions between the field and the atomic subsystem, introducing the detuning from resonance $\delta = \omega_r - \omega$, assuming it to be much smaller than the radiator frequency, ω_r.

To unify the two scattering conversions represented in **Figure 3**, it is convenient to define the number of atoms prepared in the coherent state, $|\varphi_r(0)\rangle = \exp\left[\hat{D}^-\eta\right]|j,j\rangle / \left[1 + |\eta|^2\right]^j$, where $\eta = \exp(i\varphi)\tan(\theta/2)$, $|j,j\rangle$ denotes the excited state of the dipole ensemble passing through the cavity. Following the permutation procedure under the trace operation, and working in the interaction picture, we obtain the following master equation for the cavity quasi-particles, described by the density operator,

$$\frac{\partial}{\partial t}\breve{W}(t) = (k + i\varsigma)\left[\hat{\Lambda}_n^+\breve{W}(t), \hat{\Lambda}_n^-\right] + i\chi_n 2\eta j\left[\breve{W}(t), \hat{\Lambda}_n^-\right] + H.c. \qquad (12)$$

Here, $k = 2\chi_n^2 j\gamma/(\delta^2 + \gamma^2)$ is the spontaneous conversion rate per photon, and $\varsigma = 2\chi_n^2 \delta j/(\delta^2 + \gamma^2)$ is the Raman shift from resonance stimulated by the interaction with the atomic ensemble. For small values of the polar angle θ, higher-order terms in the small parameter $|\eta| < 1.$ can be neglected.

For the small value of $|\eta|$, we can neglect the higher order terms so that for zero value of the detuning from resonance, $\delta = 0$, we obtain the following stationary equation, $k\left[\hat{\Lambda}_n^+ \breve{W}(t) + i\sigma\breve{W}(t), \hat{\Lambda}_n^-\right] + H.c, = 0$, the stationary solution of which can be represented in the normal product form, relative to the operators, $\hat{\Lambda}_n^+$, and $\hat{\Lambda}_n^-$, $\breve{W} = A\left[1/\left(\hat{\Lambda}_n^+ + i\sigma\right)\right]\left[1/\hat{\Lambda}_n^- - i\sigma\right]$. Here, $\sigma = \Omega/(2k)$, A is the normalization constant of the density matrix of the quasiparticle excitations. The behavior of the moments of the inversion operator and its quantum fluctuation are plotted in **Figure 3A–C** for the initial number of photons, $l = 8$. As follows from this representation, the fluctuations of the normal, $\Delta_{2pn}^2 = \left\langle \left(\hat{\Lambda}_2^+\right)^2 \left(\hat{\Lambda}_2^-\right)^2 \right\rangle - \left(\left\langle \hat{\Lambda}_2^+ \hat{\Lambda}_2^- \right\rangle\right)^2$, and anti-normal moments, $\Delta_{2pa}^2 = \left\langle \left(\hat{\Lambda}_2^-\right)^2 \left(\hat{\Lambda}_2^+\right)^2 \right\rangle - \left(\left\langle \hat{\Lambda}_2^- \hat{\Lambda}_2^+ \right\rangle\right)^2$, may be estimated by the stationary solutions of master equations. This square fluctuation, Δ_{2p}^2, achieved the minimal value in the region where relative conversion amplitudes the first and second steps, $g_1/g_2 = 1.4$, of coherent excitation, $\sigma = 40$.

3. Results and discussion

The Raman model demonstrates that the dimerization rate of thymine bases ($T = T$) increases exponentially with the intensity of UVC radiation, suggesting the involvement of a quantum mechanism in dimer formation. This process has been experimentally confirmed through resonant fluorescence techniques, which reveals the disruption of hydrogen bonds and associated structural transitions. In our research, we observed that membrane proteins can assemble *via* dipole–dipole interactions involving four, five, six, or more polypeptide chains, as confirmed experimentally. Although the underlying mathematical formulations are complex and not detailed here, we identified a notable rule: Planar protein structures with an odd number of polypeptide chains cannot form H-type hydrogen bonds. These packing procedures are illustrated in **Figures 4–6** using a tetramer model as an example.

We further investigated the spatial wrapping of these protein structures in three dimensions. One example involves the wrapping of equilateral triangular motifs, analyzed with π rotational symmetry along their diagonals. In these "sandwich-like" configurations, the dipoles exhibit a reduced electric moment compared to that observed within the plane of the structure. This model forms a basis for applying perturbation theory to the energy states arising from V-type dipole wrapping. We also propose analogous models for tetramers (**Figure 7**), in which the V-type dipole trapping mechanism critically influences the stability and energetic properties of the protein assemblies.

An innovative method is proposed for reconfiguring optical metamaterials composed of variable-sized quartz fibers for the disinfection of pathogenic liquids using UVC radiation. The approach involves rotating the contaminated liquid through screw channels integrated into the metamaterial of UVC spherical optical fibers, enhancing decontamination efficiency (see **Figure 10**). To validate the approach, we conducted a

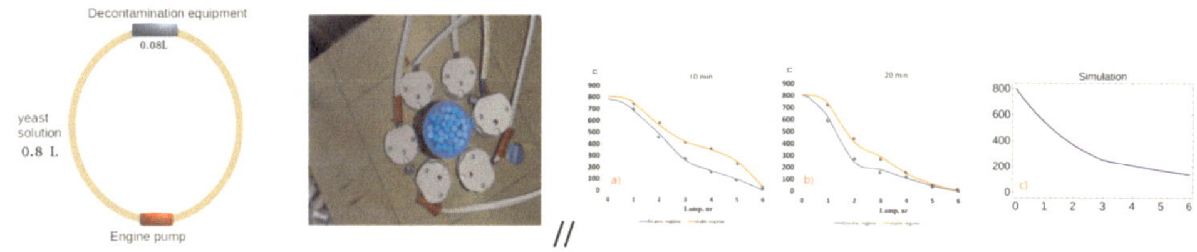

Figure 10.
Decontamination rate measured using the left-hand apparatus as a function of UVC radiation intensity, expressed in number of lamps. Panels: (a) 10-min dynamic and static regimes; (b) 20-min dynamic and static regimes; (c) dependence of the inactivation rate on applied intensity, as predicted by the proposed theoretical model [46].

study of dynamic and static inactivation of bakery yeast using UV regimes. In the static test, we used 80 ml of fungus solution, setting a short decontamination time. The yeast solution was prepared from 25 g of yeast dissolved in 800 *ml* of water and 20 g of agar ($C14H24O9$) to support the development of fungus. At the initial stage, we observed the formation of variable yeast colonies. Previous experiments have demonstrated their inactivation in 15 minutes, and in this phase, we achieved inactivation after 24 hours of growth. We investigated the impact of UVC light intensity on the inactivation rate, adding up to six UVC lamps to the decontamination core, according to **Figure 10**.

4. Conclusions

The proposed methods demonstrate that proteins and biomolecules can be repackaged into supramolecular structures through dipolar stimulation *via* multiple Raman photon conversions. This process enables both the synthesis and breakdown of molecular assemblies by creating superpositions between ground and excited states with nonzero dipole momentum.

Quantum cooperation within Raman scattering allows controlled formation of aggregated biomolecules, including multimers, and facilitates the efficient transfer of excitation or information between conversion stages. Cooperative phenomena, such as the Dicke effect, play a key role in stabilizing dimerized and multimerized states, particularly in nucleic acids and protein substructures.

Experimentally, these mechanisms can generate excitons in supramolecular structures and can be exploited for pathogen inactivation. In particular, the dimerization and multimerization of DNA and protein substructures provide a promising approach for improving UVC-based decontamination. Overall, this study highlights the potential of combining Raman-mediated dipolar stimulation with UVC optical metamaterials to manipulate biomolecular assemblies and enhance disinfection efficiency.

Acknowledgements

The authors gratefully acknowledge the scientific and material support provided by the Institute Project "Quantum Cooperative Phenomena, 011206," the State Project "Stimulating Excellence" in Research for the Years 2024–2025, 24.80012.5007.26SE and 25.80012.5007.36TC.

The author acknowledges the use of ChatGPT (OpenAI) for language polishing of the manuscript.

Author details

Nicolae A. Enaki*, Diana Podoleanu, Ion Munteanu, Gheorghe Chirman,
Elena Starodub and Tatiana Paslari
Quantum Optics and Kinetic Processes Lab, Institute of Applied Physics of Moldova State University, Chisinau, Republic of Moldova

*Address all correspondence to: enakinicolae@yahoo.com

References

[1] Blombergen N, Shen YR. Quantum-theoretical comparison of nonlinear susceptibilities in parametric media, lasers, and Raman lasers. Physics Review. 1964;**133**:A37-A49

[2] Shen YR, Blombergen N. Theory of stimulated Brillouin and Raman scattering. Physics Review. 1965;**137**: A1787-A1805

[3] Wang C-S. Theory of Stimulatel Raman scattering. Physics Review. 1969; **182**:482-494

[4] Chen YF, Chen YS, Tsai SW. Diode-pumped Q-switched laser with intracavity sum frequency mixing in periodically poled KTP. Applied Physics B: Lasers and Optics. 2004;**79**(2): 207-210

[5] Jia Z, Li J, Gao L, Yang D, Kanaev A. Dynamic light scattering: A powerful tool for In situ nanoparticle sizing. Colloids and Interfaces. 2023;7(1):15. DOI: 10.3390/colloids7010015

[6] Li F, Wang X, Zhao X, et al. First-stokes and second-stokes multi-wavelength continuous-wave operation in Nd:YVO4/BaWO4 Raman laser under in-band pumping. Chinese Optics Letters. 2020;**8**(11):111401

[7] Chencheng Shen C, Xianglong CX, Sang Y, et al. Investigation of multispectral SF6 stimulated Raman scattering laser. Chinese Optics Letters. 2020;**18**(5):051402. DOI: 10.3788/COL202018.051402

[8] Anfinsen CB. Principles that govern the folding of protein chains. Science. 1973;**181**(4096):223-230. DOI: 10.1126/science.181.4096.223

[9] Richardson JS. Handedness of crossover connections in b-sheets. Proceedings of the National Academy of Sciences of the United States of America. 1976;**173**:2619-2623

[10] Cole BJ, Bystroff C. Alpha helical crossovers favor right-handed supersecondary structures by kinetic trapping: the phone cord effect in protein folding. Protein Science. 2009;**18** (8):1602-1608. DOI: 10.1002/pro.182

[11] Takano Y, Kondo HX, Nakamura H. Quantum chemical studies on hydrogen bonds in helical secondary structures. Biophysical Reviews. 2022;**14**:1369-1378. DOI: 10.1007/s12551-022-01034-5

[12] Celardo GL, Angeli M, Craddock TJA, et al. On the existence of superradiant excitonic states in microtubules. New Journal of Physics. 2019;**21**:023005. DOI: 10.1088/1367-2630/aaf839

[13] Hestand NJ, Spano FC. Molecular aggregate photophysics beyond the Kasha model: Novel design principles for organic materials. Accounts of Chemical Research. 2017;**50**(2):341-350. DOI: 10.1021/acs.accounts.6b00576

[14] Kasha M. Energy transfer mechanisms and the molecular exciton model for molecular aggregates. Radiation Research. 1963;**20**:55-70

[15] Hochstrasser RM, Kasha M. Application of the exciton model to monomolecular lamellar systems. Photochemistry and Photobiology. 1964; **3**:317-331

[16] Enaki NA. Two types of cooperation between the photons in multiple scattering lasing and their possible application. Physics Open. 2024;**21**: 100231. DOI: 10.1016/j.physo.2024.100231

[17] Enaki NA. Cooperative properties of multiple quantum scattering: I quantum nutation. Physica Scripta. 2024;**99**(4): 045102. DOI: 10.1088/1402-4896/ad29cb

[18] Enaki NA. Cooperative properties of multiple quantum scattering: II. Coherentlasing. Physica Scripta. 2024;**99**(4):045101. DOI: 1088/1402-4896/ad2ac0

[19] Enaki NA. Mutual cooperative effects between the mode components of two-photon and Raman induced cavity lasing processes. Optics Communication. 2021;**498**:127124. DOI: 10.1016/j.optcom.2021.127124

[20] Kapteyn HC, Murnane MM, Christov IP. Extreme nonlinear optics: Coherent X rays from lasers. Physics Today. 2005;**58**(3):39-46. DOI: 10.1063/1.1897563

[21] Hentschel M, Kienberger R, Spielmann C, et al. Attosecond metrology. Nature. 2001;**414**:509-513. DOI: 10.1038/35107000

[22] Paul PM, Toma ES, Bregen P, et al. Observation of a train of attosecond pulses from high harmonic generation. Science. 2001;**292**:1689-1692. DOI: 10.1126/science.1059413

[23] Tzallas P, Charalambidis D, Papadogiannis NA, et al. Direct observation of attosecond light bunching. Nature. 2003;**426**:267-271. DOI: 10.1038/nature02091

[24] A. Baltuska A., Udem T., Uiberacker M. et al. Attosecond control of electronic processes by intense light fields. Nature. 2003;**421**:611. DOI: 10.1038/nature01414

[25] Harris SE, Sokolov AV. Subfemtosecond pulse generation by molecular modulation. Physical Review Letters. 1998;**81**:2894. DOI: 10.1103/PhysRevLett.81.2894

[26] Nazarkin A, Korn G, Wittmann M, Elsaesser T. Generation of multiple phase-locked stokes and anti-stokes components in an impulsively excited Raman medium. Physical Review Letters. 1999;**83**:2560. DOI: 10.1103/PhysRevLett.83.2560

[27] Sali E, Kinsler P, New GHC, et al. Behavior of high-order stimulated Raman scattering in a highly transient regime. Physical Review A. 2005;**72**: 013813. DOI: 10.1103/PhysRevA.72.013813

[28] Nazarkin A, Korn G, Elsaesser T. All-linear control of attosecond pulse generation. Optics Communication. 2002;**203**:403-412. DOI: 10.1016/S0030-4018(02)01133-1

[29] Schreier WJ, Schrader WJ, Koller FO, et al. Thymine dimerization in DNA is an ultrafast photoreaction. Science. 2007;**315**:625-629. DOI: 10.1126/science.1135428

[30] Boixeda P, Carmona LP, Vano-Galvan S, et al. Advances in treatment of cutaneous and subcutaneous vascular anomalies by pulsed dual wavelength 595- and 1064-nm application. Medical Laser Application. 2008;**23**(3):121-126. DOI: 10.1016/j.mla.2008.04.002

[31] Enaki NA, Paslari T, Bazgan S, et al. UVC radiation intensity dependence of pathogen decontamination rate: Semiclassical theory and experiment. European Physical Journal Plus. 2022;**137**(9):1047-1-1047-14. DOI: 10.1140/epjp/s13360-022-03252-y

[32] Riazi A, Chen C, Zhu EY, et al. Biphoton shaping with cascaded entangled-photon sources. npj Quantum

Information. 2019;**5**:77. DOI: 10.1038/s41534-019-0188-1

[33] Khan U, Kasha M. Mechanism of four-level laser action in solution excimer and excited-state proton-transfer cases. Proceedings of the National Academy of Sciences of the United States of America. 1983;**80**: 1767-1770

[34] Ojeda-May P, Garcia ME. Electric field-driven disruption of a native b-sheet protein conformation and generation of a helix-structure. Biophysical Journal. 2010;**99**:595-599. DOI: 10.1016/j.bpj.2010.04.040

[35] Callaway E. "A Pandora's box": map of protein-structure families delights scientists. Nature. 2023;**621**(7979):455. DOI: 10.1038/d41586-023-02892-z

[36] Chiang YS, Gelfand TI, Kister AE, Gelfand IM. New classification of supersecondary structures of sandwich-like proteins uncovers strict patterns of strand assemblage. Proteins. 2007;**68**(4): 915-921. DOI: 10.1002/prot.21473

[37] Kasha M, Rawls HR, El-Bayoumi MA. The exciton model in molecular. Spectroscopy. 1965;**11**:371-392

[38] He J, Hu HJ, Harrison R, Tai PC, Pan Y. Rule generation for pro- tein secondary structure prediction with support vector machines and decision tree. IEEE Transactions on Nanobioscience. 2006;**5**:46-53

[39] Aydin Z, Altunbasak Y, Borodovsky M. Protein secondary structure prediction for a single-sequence using hidden semi-Markov models. BMC Bioinformatics. 2006;**7**:178

[40] Lin K, Simossis VA, Taylor WR, Heringa J. A simple and fast secondary structure prediction method using hidden neural net- works. Bioinformatics. 2005;**21**(152–159):5

[41] Crooks GE, Brenner SE. Protein secondary structure: entropy, correlations and prediction. Bioinformatics. 2004;**20**:1603-1611

[42] Pollastri G, McLysaght A. Porter: a new, accurate server for protein secondary structure prediction. Bioinformatics. 2005;**21**:1719-1720. DOI: 10.1093/bioinformatics/bti203

[43] Zhao L, Li J, Zhan W, et al. Prediction of protein secondary structure by the improved TCN-BiLSTM-MHA model with knowledge distillation. Scientific Reports. 2024;**14**: 16488. DOI: 10.1038/s41598-024-67403-0

[44] Zakharov OS, Rudik AV, Filimonov DA, Lagunin AA. Prediction of protein secondary structures based on substructural descriptors of molecular fragments. International Journal of Molecular Sciences. 2024;**25**:12525. DOI: 10.3390/ijms252312525

[45] Vignesh U, Parvathi R, Gokul Ram K. Ensemble deep learning model for protein secondary structure prediction using NLP metrics and explainable AI. Results in Engineering. 2024;**24**:103435. DOI: 10.1016/j.rineng.2024.103435. ISSN 2590-1230

[46] Munteanu I, Starodub E, Bazgan S, Turcan M, Paslari T, Podoleanu D, et al. Ultraviolet C intensity dependence of decontamination efficiency for pathogens as function of repacked metamaterials with screw channels. European Biophysics Journal. 2024;**53** (3):133-145. DOI: 10.1007/s00249-024-01702-2

[47] Wang H, Zhang L, Huang J, Yang Z, Fan C, Yuan L, et al. Imaging the

intracellular refractive index distribution (IRID) for dynamic label-free living colon cancer cells via circularly depolarization decay model (CDDM). Biomedical Optics Express. 2024;**15**(4): 2451-2465. DOI: 10.1364/BOE. 518957

[48] Chamot S, Migacheva E, Seydoux O, et al. Physical interpretation of the phase function related parameter γ studied with a fractal distribution of spherical scatterers. Optics Express. 2010;**18**: 23664-23675. DOI: 10.1364/ OE.18.023664

[49] Brunsting A, Mullaney PF. Differential light-scattering from spherical mammalian-cells. Biophysical Journal. 1974;**14**:439-453. DOI: 10.1016/ S0006-3495(74)85925-4

[50] Maier JS, Walker SA, Fantini S, et al. Possible correlation between blood-glucose concentration and the reduced scattering coefficient of tissues in the near-infrared. Optics Letters. 1994;**19**: 2062-2064. DOI: 10.1364/ol.19.002062

[51] Liu H, Beauvoit B, Kimura M, Chance B. Dependence of tissue optical properties on solute-induced changes in refractive index and osmolarity. Journal of Biomedical Optics. 1996;**1**(2):200-211. DOI: 10.1117/12.231370

[52] Kurian P, Dunston G, Lindesay J. How quantum entanglement in DNA synchronizes double-strand breakage by type II restriction endonucleases. Journal of Theoretical Biology. 2016;**391**:102-112. DOI: 10.1016/j.jtbi.2015.11.018

[53] Mukherjee S, Majumdar S, Bhattacharyya D. Role of hydrogen bonds in protein-DNA recognition: effect of nonplanar amino groups. The Journal of Physical Chemistry. B. 2005; **109**(20):10484-10492. DOI: 10.1021/ jp0446231

[54] Singh V, Fedeles BI, Essigmann JM. Role of tautomerism in RNA biochemistry. RNA. 2015;**21**(1):1-13. DOI: 10.1261/rna.048371.114

[55] Slocombe L, Sacchi M, Al-Khalili J. An open quantum systems approach to proton tunnelling in DNA. Communications on Physics. 2022;**5**:109. DOI: 10.1038/s42005-022-00881-8

[56] Watson JD, Crick FHC. The structure of DNA. Cold Spring Harbor Symposia on Quantitative Biology. 1953; **18**:123-131. Available from: https://api. semanticscholar.org/CorpusID:16868783

[57] Alberts BB. "Chromosomal DNA and its Packaging in the Chromatin Fiber". Molecular Biology of the Cell. 4th ed. New York: Garland Science; 2002. p. 207. ISBN 978-0-8153-4072-0.

[58] Youngson RM. Collins Dictionary of Human Biology. Glasgow: Harper Collins; 2006. ISBN 978-0-00-722134-9

[59] Cox M, Nelson DR, Lehninger AL. Lehninger Principles of Biochemistry. San Francisco: W.H. Freeman; 2005. ISBN 978-0-7167-4339-2

[60] Redon C, Pilch D, Rogakou E, et al. Histone H2A variants H2AX and H2AZ. Current Opinion in Genetics & Development. 2002;**12**(2):162-169. DOI: 10.1016/S0959-437X(02)00282-4

[61] dos Santos CS, Drigo FE, Ricotta RM. Quantum confinement in hydrogen bond. Journal of Physics: Conference Series. 2015;**597**:12033. DOI: 10.1088/ 1742-6596/597/1/012033

[62] Godbeer AD, Al-Khalili JS, Stevenson PD. Environment-induced dephasing versus von Neumann measurements in proton tunneling. Physical Review A. 2014;**90**:012102. DOI: 10.1103/RevModPhys.35.724

[63] Enaki NA. Non-linear Cooperative Effects in Open Quantum Systems: Entanglement and Second Order Coherence. NY: Nova Science Publisher; 2016, 344 p

[64] Per-Olov L. Proton tunneling in DNA and its biological implications. Reviews of Modern Physics. 1963;**35**:724-732. DOI: 10.1103/RevModPhys.35.72

[65] Bhagavan NV, Ha C-E. Structure and properties of DNA. In: Essentials of Medical Biochemistry. London, UK: Elsevier; 2015. pp. 381-400

[66] Puri RR, Lawande SV. Exact steady-state density operator for a collective atomic system in an external field. Physics Letters A. 1979;**72**:200-202

[67] Puri RR, Lawande SV, Hassan SS. Dispersion in the driven Dicke model. Optics Communication. 1980;**35**:179-184. DOI: 10.1016/0030-4018(80) 90205-9

[68] Drummond PD, Walls DF. Quantum theory of optical bistability. I: Nonlinear polarisability model. Journal of Physics A. 1980;**13**:725. DOI: 10.1088/0305-4470/13/2/034